GULF OIL AND GAS

ENSURING ECONOMIC SECURITY

GULF OIL AND GAS

ENSURING ECONOMIC SECURITY

THE EMIRATES CENTER FOR STRATEGIC
STUDIES AND RESEARCH

THE EMIRATES CENTER FOR STRATEGIC STUDIES AND RESEARCH

The Emirates Center for Strategic Studies and Research (ECSSR) is an independent research institution dedicated to the promotion of professional studies and educational excellence in the UAE, the Gulf and the Arab world. Since its establishment in Abu Dhabi in 1994, the ECSSR has served as a focal point for scholarship on political, economic and social matters. Indeed, the ECSSR is at the forefront of analysis and commentary on Arab affairs.

The Center seeks to provide a forum for the scholarly exchange of ideas by hosting conferences and symposia, organizing workshops, sponsoring a lecture series and publishing original and translated books and research papers. The ECSSR also has an active fellowship and grant program for the writing of scholarly books and for the translation into Arabic of work relevant to the Center's mission. Moreover, the ECSSR has a large library including rare and specialized holdings, and a state-of-the-art technology center, which has developed an award-winning website that is a unique and comprehensive source of information on the Gulf.

Through these and other activities, the ECSSR aspires to engage in mutually beneficial professional endeavors with comparable institutions worldwide, and to contribute to the general educational and academic development of the UAE.

The views expressed in this book do not necessarily reflect those of the ECSSR.

First published in 2007 by
The Emirates Center for Strategic Studies and Research
PO Box 4567, Abu Dhabi, United Arab Emirates

E-mail: pubdis@ecssr.ae
Website: http://www.ecssr.ae

Distributed by The Emirates Center for Strategic Studies and Research

ISBN: 978-9948-00-856-9 hardback edition
ISBN: 978-9948-00-857-6 paperback edition

Contents

Gulf Oil and Supply Security

Future Demand for Gulf Oil

Figures and Tables

Figures

TABLES

Abbreviations and Acronyms

AEO	Annual Energy Outlook
AMD	Advanced Micro Devices
API	American Petroleum Institute
bcf	billion cubic feet
bpd	barrels per day
BTU	British Thermal Units
CEO	Chief Executive Officer
CERA	Cambridge Energy Research Associates
CGES	Centre for Global Energy Studies
CIS	Commonwealth of Independent States
CNBS	China National Bureau of Statistics
CNPC	China National Petroleum Corporation
CPI	Consumer Price Index
CSIS	Center for Strategic and International Studies
CTL	coal-to-liquids
DOE	Department of Energy
E&P	exploration and production
EIA	Energy Information Administration
FDI	foreign direct investment
FERC	Federal Energy Regulatory Commission
FGE	FACTS Global Energy
FSU	Former Soviet Union
GCC	Gulf Cooperation Council
GDP	gross domestic product
GTL	gas-to-liquids
GW	gigawatt
HETCO	Hess Energy Trading Company

IEA	International Energy Agency
IEO	International Energy Outlook
IFP	Institut Français du Pétrole
IIF	Institute for International Finance
IMF	International Monetary Fund
IPE	International Petroleum Exchange
IOC	international oil company
IPO	initial public offering
IPEC	Independent Petroleum Exporting Countries
IRAC	imported refiner acquisition cost
KPC	Kuwait Petroleum Corporation
LPG	liquefied petroleum gas
LNG	liquefied natural gas
mbpd	million barrels per day
MEC	Middle East Consultants
MENA	Middle East and North Africa
MIT	Massachusetts Institute of Technology
MOI	Ministry of Interior
NEMS	National Energy Modeling System
NGL	natural gas liquid
NIOC	National Iranian Oil Company
NOC	national oil company
NYMEX	New York Mercantile Exchange
OECD	Organization for Economic Cooperation and Development
OIAF	Office of Integrated Analysis and Forecasting
OPEC	Organization of Petroleum Exporting Countries
PEL	Petroleum Economics Ltd.
PCEC	primary commercial energy consumption
PIRA	Petroleum Industry Research Associates

PISF	Petroleum Installation Security Force
PIW	Petroleum Intelligence Weekly
RES	Reference Energy System
R/P	Reserves-to-Production (ratio)
SAGE	System for the Analysis of Global Energy Markets
SUV	Sport Utility Vehicle
TFP	total factor productivity
TRC	Texas Railroad Commission
UAE	United Arab Emirates
USGS	United States Geological Survey
WTI	West Texas Intermediate
WTO	World Trade Organization

FOREWORD

In recent years, the world has witnessed a situation of tight oil supply and soaring oil prices, which has enabled the energy producers of the Arabian Gulf to capitalize on massive petroleum revenues, with favorable effects on their national budgets and economies. However, this revenue windfall has coincided with heightened political and strategic uncertainty in the region. It also comes at a time when the main energy producers, led by OPEC, are facing greater challenges in increasing oil production capacity, ensuring secure oil supplies, dealing with constraints in crude quality and refining capacity and regulating the oil market in the face of burgeoning demand.

The ECSSR 11th Annual Energy Conference entitled *Gulf Oil and Gas: Ensuring Economic Security* was held from September 25 to 27, 2005 in Abu Dhabi. The conference brought together distinguished scholars, senior economists and energy experts from different parts of the world. It aimed to assess global energy trends, sketch their implications and pinpoint strategies for the Gulf states to strengthen their economic security. Participants outlined the economic impact of record-high oil prices, projecting both the long-term benefits and challenges for the Gulf oil producers. The conference examined the potentially destabilizing effect of geopolitical crises on energy markets; analyzed energy trends, such as the globalization of the gas trade, and how crude quality trends and demand for different grades of oil would affect energy markets. The forum also examined avenues for cooperation between consumers and producers, and between national and foreign oil companies, focusing particularly on the investment prospects for the latter in the Gulf region. The projected demand for Gulf energy and future prospects for oil production were assessed with particular focus on Saudi Arabia's sustainable production capacity and oilfield security concerns.

The conference highlighted the fact that cooperation between energy exporters and energy importers is vital in balancing the opposing needs of supply and demand security and supporting future global economic growth. With world energy demand set to surge considerably, especially in Asia, massive international investment was viewed as imperative, not only to expand Gulf oil and gas production but also to create additional oil refining capacity to deal with heavier crude categories. There was broad consensus that surplus oil revenues afforded an exceptional opportunity for the Gulf states to meet their current and future development challenges by boosting energy investment levels, promoting economic diversification, embarking on new economic ventures and initiating social programs to lay the foundation for the region's future economic security. The forum stressed that for the Gulf states, the concept of economic security must extend well beyond petroleum resource management to encompass pro-active policies and plans that could propel the region towards long-term diversified development and greater regional cohesion and ensure its strategic position in an increasingly integrated world economy.

In accordance with the ECSSR's mission, the informative conference presentations have been compiled in this volume for the benefit of researchers, policy makers and professionals working in the field of energy. The ECSSR would like to express its appreciation to all the conference speakers for their active participation and for sharing their expertise with readers through this volume. The Center is indebted to the referee panel of ten energy experts who offered their valuable comments and recommendations on the conference papers. A word of thanks is also due to ECSSR editor Mary Abraham for coordinating the publication of this book.

Jamal S. Al-Suwaidi, Ph.D.
Director General
ECSSR

INTRODUCTION

Economic Security Issues and the Role of the Gulf Energy Sector

In recent years, the world has witnessed a situation of high oil prices stemming from a combination of factors: geopolitical disruptions, tight oil supply, refining constraints and relentless global energy demand led by rapidly growing Asian economies. The unprecedented oil revenues enjoyed by the Arabian Gulf oil producers have presented them with a unique opportunity to restructure, reform, diversify and integrate their economies. However, formidable challenges remain for the Gulf producers as they seek to meet global demand at acceptable prices while judiciously managing their oil and gas reserves and ensuring long term economic security.

The ECSSR 11th Annual Energy Conference, *Gulf Oil and Gas: Ensuring Economic Security,* held from September 25–27, 2005 in Abu Dhabi, brought together an international panel of energy experts and analysts to discuss crucial energy developments, assess industry trends and outline effective ways for the Arabian Gulf producers to fortify their economic security in the coming decades. The topics discussed included long-term production and pricing strategies for OPEC; the changing concept of Gulf economic security; GCC strategies to overcome the "resource curse;" deterioration in future crude quality; capacity constraints and emerging refining challenges; global energy investment requirements; Saudi Arabia's reserves and sustainable production capacity; supply security concerns of major consuming nations; Asian and US energy import dependence on the Gulf; ensuring demand security for oil producing nations and the shifting importance of various energy sources in the global energy slate. The conference participants sought to address these important energy issues, reaching significant conclusions and offering noteworthy recommendations in their presentations, which are summarized below.

Toward Gulf Economic Security

In contrast to past oil shocks, the current high oil prices are the outcome of global economic recovery, which has resulted in high levels of oil consumption and demand. This increase in demand has cut into spare global production capacity, further heightening oil price volatility. In his keynote address, H.E. Sheikh Ahmad Fahad Al Ahmad Al Sabah observes that the world economy has shown increasing resilience in enduring the impact of higher prices. However, the negative effects of a volatile oil market on global economic security in general and oil producing economies in particular should not be underestimated. OPEC comprehends these effects well and recognizes that they ultimately harm both the consuming and the producing nations. The producing nations are considered the sole beneficiaries of high oil prices and are often perceived as having orchestrated them. Not surprisingly, reports frequently view OPEC as being responsible for any rise in oil prices, resulting in repeated calls to member states to urgently raise their production levels.

High oil prices that undermine the economies of consuming nations do not serve the ultimate interests of the producing nations, especially OPEC members. Past experience suggests that such high prices cannot be sustained and are bound to collapse. It is therefore imperative for OPEC to maintain a robust world oil demand by upholding a reasonable price level. Keeping this in mind, OPEC has made every attempt possible to lower the current prices. It has not only raised its production ceiling several times during the current price rise but is prepared to raise it even further whenever market stability is at risk. OPEC members have also pumped up production to nearly full capacity in order to meet the anticipated demand.

Price-related Challenges for Producing Nations

High oil prices bring immediate revenue benefits to the producing nations but they also pose some particular problems. Herman T. Franssen pinpoints the causative factors behind the high prices and outlines the main challenges that the oil producing nations will have to tackle in the short, medium and long terms.

In the short term, the crude oil market seems well supplied, as indicated by the high level of OECD commercial stocks. The main constraints are to be found in the refining sector, where upgrading capacity that can convert high sulfur, heavy crude oil into light, low-sulfur products, has fallen short of demand. Concerns over oil supply disruptions, particularly in Africa and the Middle East, have also helped to boost oil prices. OPEC is producing the maximum quantity of crude oil that the global refining system can handle and several member states are expanding their export refining capacity.

In the medium term, additional non-OPEC and OPEC capacity growth is expected to provide several million barrels per day of spare crude oil capacity. This would allow OPEC to manage supply better and seek to maintain a desirable oil price level. There is concern that OECD government policies to reduce oil imports from the Middle East will adversely impact demand when massive new upstream investments are being made by OPEC members to meet the projected global demand. Hence, oil producers are looking for guarantees to ensure "oil demand security" just as consumer countries are pushing to gain oil supply security.

Global upgrading capacity in refining will take years to match the projected levels of global oil demand growth and product prices may therefore continue to reflect this tight refining situation. OPEC plans to ease this bottleneck by adding massive additional refining capacity, particularly in the Arabian Gulf countries.

In the long run, the demand for transportation fuels in the emerging economies of Asia, the Middle East and Latin America will pose a major challenge for OPEC, particularly for the high reserve, low consumption countries of the Arabian Gulf. This pent-up long term transportation fuel demand is truly staggering and there are no available alternatives. While non-conventional fuels (hydrocarbon and bio-fuels) will provide some supplies at the margin, oil will remain the mainstay of the transportation sector for decades to come. The main challenge for the Arab Gulf states is to manage their oil reserves so as to fulfill global oil demand at "reasonable" prices, while preserving much of their resources for future generations.

Globalization and Gulf Economic Security

The current Gulf economic boom has been the cumulative outcome of rising oil revenues, expansionary fiscal policies and massive liquidity flowing through the financial markets. If oil prices continue to remain at high levels, the Gulf economies can sustain this newly found prosperity but there lies the real danger. Tarik M. Yousef warns that such exceptional conditions are unlikely to last and may divert attention from important policy objectives such as long-term economic growth and stability. The pursuit of these objectives should guide any conceptualization of Gulf economic security in the 21st century.

Future Gulf economic security demands not only global economic integration and proactive strategies to manage its risks, but also the integration of this agenda into broader national development strategies. Otherwise, countries run the risk of amplifying the costs of adjustment and dislocation that may arise, not to mention the neglect of other domestic policy priorities. These two policy objectives – globalization and national development – are not only potentially complementary but also mutually reinforcing.

The persistent and rising unemployment levels over the past decade along with the projected labor market flows has raised doubts about the merits of selective reform programs typically adopted by Middle Eastern countries. Whether in relaxing labor market regulations, downsizing the public sector or nationalizing their labor forces, the record of the last decade suggests that direct interventions in Gulf labor markets alone will be insufficient. Other selective measures such as the privatization of public sector enterprises, establishment of free trade zones or launch of international joint ventures have also proved inadequate. In isolation, these measures do not constitute an effective development strategy for the Gulf countries.

Emerging conventional wisdom urges the Gulf countries to address a set of long-standing policy and institutional challenges to complete three fundamental, interrelated realignments within their economies:

- From public sector to private sector dominated economies, by reducing the barriers to private activity while creating regulatory frameworks that ensure that private and social interests coincide.
- From closed to more open economies, by facilitating integration into global commodity and factor markets while establishing safeguards for financial stability and social protection.
- From oil-dominated and volatile to more stable and diversified economies, by making fundamental changes in institutions managing oil resources and their intermediation to economic agents.

Moreover, for the region to realize this long-standing transition, rapid progress is needed in educational reform, gender equality and better governance. In other words, the new wisdom urges a rapid implementation of the broad, unfinished reform agenda initiated in the early 1990s.

The GCC: Overcoming the Resource Curse

Economic restructuring and reform is imperative, not only to integrate and diversify the economy but also to help avoid the problems inherent in oil-dependent economies. The current global oil boom provides a historic opportunity for the Gulf Cooperation Council states to overcome the "resource curse," which has afflicted their economies and societies over the past half century. Unlike the oil boom of the 1970s, the GCC countries now have the human resource base and absorptive capacity needed to lay the foundation for sustained economic growth when the current expansion phase of the petroleum sector ends and the economic cycle moves toward its next trough. However, as Edward L. Morse points out, there are powerful impediments to the kind of reforms needed for the governments in the region to overcome their endemic resource curse.

It is an opportune moment for GCC governments to rethink the strategy of having huge state companies operating in the oil sector. This is no easy task in a region where the form of government is monarchical rather than participatory, and where the state revenues and the dispensation of essentially free state services have traditionally boosted governmental authority and legitimacy. Undoubtedly there is pent-up demand within the GCC for direct participation in the oil sector and partial privatizations could go a long way to slaking this desire. Certainly, with enhanced property rights and the protective rule of law in defending those rights, the flourishing local capital markets can be used to spread that wealth.

The very idea of dismantling these key state assets or doing so too quickly will definitely lead to a chorus of objections. However, internal GCC debates should lead to a consensus on the appropriate path to be followed. Dubai's effective transformation from an oil-producing emirate to a diversified services economy suggests one workable path toward economic diversification. The successful privatizations of once large state-owned enterprises in oil, gas and other energy sectors in the OECD countries also demonstrates that significant benefits can be earned both by state and society by changing the property rights system governing the oil and gas sectors.

In this connection, the Norwegian experience is quite instructive. In the process of privatization, the Norwegian government discovered that it has lost little control over the oil sector – indeed in many respects it acquired some of the authority it lacked when it had a huge national oil company – and gained much by separating the regulatory functions over oil and gas operations from the actual ownership. It has gained from more competition and greater participation by citizens in oil and gas property rights. However, it can still act to curb production if prices fall suddenly and damage the domestic economy.

It may be concluded that in the current climate of oil sector expansion, there are opportunities to seize win–win solutions. This is especially true of the smaller GCC states, which can integrate their economies into global market mechanisms with much greater ease than the larger GCC members.

Matching Crude Oil Quality to Clean Product Demand

In the context of oil sector expansion, new opportunities have brought fresh challenges in their wake. Important changes have occurred on both the supply and demand sides that have led to great tightness in the markets. In addition, the growing mismatch between deteriorating crude quality and rising clean product demand is creating a new refining challenge, which may be summarized in one sentence: "The last barrel produced will be medium/heavy and sour, while the last barrel demanded will be light (clean)." As Pedro Antonio Merino Garcia observes, there is one way to balance crude supply and final product demand requirements and that is to invest in more capacity in the refining industry.

Effectively, the demand side is characterized by an increasing push for clean products. Demand is being driven by strong economic growth, the growing participation of the transport sector in total final consumption and by the challenge posed by new ecological considerations. This demand resilience stems from the fact that, both now and in the medium term, there is no effective substitute for gasoline and diesel to fuel the global vehicular motion. With regard to the new environmental restrictions, although not all regions start from the same experience level, all have passed legislation to further reduce sulfur content in final oil products.

On the supply side, no analyst expects any real improvement in oil quality. No more than 60 percent of total oil production will continue to be medium and heavy for the foreseeable future. Moreover, if IEA long term forecasts of oil quality are considered or implicit projections are done on the basis of long term forecasts of OPEC market share, then the expected oil quality will further deteriorate.

Even the most optimistic observers predict a significant rise of sulfur content in the average global supply. Crude oil quality is evidencing a quality deterioration which, even if it is considered "slight," could nevertheless denote a significant challenge for a refining industry without adequate conversion (upgrading) capacity to meet the clean demand requirements.

A joint projection of the demand growth rates for clean products and the increases in refining capacity for such products, suggests that supply will not meet demand until 2013. This imbalance can only be resolved by investing in refining (conversion units), but investment should be based on a high margin outlook. In this respect, some factors discouraging investment are that refining margins have been historically low and volatile, and there is a systematic difference between crude oil price and refining margin forecast and current data.

Considering that the need for OPEC crude is expected to increase and its petroleum quality is medium/heavy and sour, crude oil price differentials will remain high and investing in conversion units should remain profitable. Nevertheless, the timing for investment in deep conversion units can be crucial. Moreover, a substantial part of investment must go towards meeting the more stringent requirements in terms of sulfur content. This investment will divert funding from conversion units, diminishing the risk of overinvestment in conversion and will contribute to a tight refining situation by reducing refinery yields.

Investment Risks and Potential for Foreign Oil Companies:

The refining sector is not the only one requiring heavy investment. With rising oil consumption reducing existing capacity surpluses, considerable investment is needed also in the oil production and transport sectors. While urging a detailed assessment of investment requirements, Jean-Pierre Favennec points out that the Gulf countries, where two-thirds of the world's oil reserves are located, will undoubtedly have a decisive role to play.

The IEA predicts that, between 2000 and 2030, approximately three-fourths of additional global oil supply will originate from the Middle East. The region would then represent 40 percent of the world's oil supply. Thus the challenge would be to increase Middle East production from 27 to around 45 million barrels per day between 2005 and 2030, which implies building a total production capacity (additional capacity as well as replacement of depleted capacities) of over 40 million barrels per day. Gas exports, on the other hand, should increase tenfold during the same period and the Middle East will become the top exporter of natural gas.

For the period from 2003 to 2030, the IEA forecasts an investment need of US$16 trillion, which amounts to US$568 billion per year, in the energy sector worldwide. More than 60 percent of these investments will be earmarked for electricity and 18 percent each for oil and gas. An amount of US$100 billion per year will be invested in the oil sector.

The Middle East region alone will need to invest a total of US$1 trillion between 2000 and 2030 for all energy sources—including US$500 billion for oil and US$200 billion for gas. This is the area with the biggest investments in hydrocarbons production, given the size of the reserves and the urgent need to expand production to satisfy global energy requirements.

However, the general reluctance of producer countries to open their upstream sector to foreign companies remains a major hurdle. This reluctance stems first, from their desire to maintain control of resources that are considered as public property. Second, the OPEC quota system tends to impose some production limits. Third, these states currently enjoy large budget surpluses, and have no immediate incentive to attract foreign companies.

Large international oil companies hesitate to invest for other reasons. In the medium term, these companies anticipate a barrel price much lower than the current price (US$25–30 rather than US$50) and expect tax system revisions to skim off the surplus revenues. The most promising energy reserves are not made available for extraction, leading to assertions that international companies "lack profitable projects." Where profitable projects do exist, the conditions are considered too risky. Social instability in Saudi Arabia, the frequent attacks in Iraq and the uncertainty over the future status of contracts, and Iran's international political situation are contexts that do not encourage investment. The political stabilization of the oil-producing regions and the promotion of transparency, good governance and a clear regulatory framework are therefore necessary conditions for international oil companies to expand investment. Although these companies want to find avenues for investment in the Arabian Gulf, with its considerable hydrocarbon wealth, the current climate is not particularly conducive.

Saudi Arabia: Raising Sustainable Capacity and Oil Security

When it comes to existing oil production capacity and future potential for expansion, the Kingdom of Saudi Arabia remains the key player. The Kingdom not only holds the largest oil reserves and is the largest oil producer in the world, but is also a highly influential OPEC member. Moreover, Saudi Arabia's spare oil production capacity assumes great importance to the global market as there is hardly any substantial spare capacity outside the Kingdom today, especially in such a tight market situation.

Nawaf E. Obaid reveals that the Saudi production program aims to increase the country's sustainable capacity substantially over five years from 2004 to 2009. Over this period, Haradh, Khursaniyah, Shaybah, Nuayyim and Khoreis are the five fields that would bring the incremental increase to the final tally of 12.5–12.6 million barrels of sustained capacity by 2009. Meanwhile the Saudi refining expansion program should bring the current capacity from 3.9 million barrels per day to a target of approximately 6 million bpd, hopefully by 2011–2012.

The Saudi security budget has expanded and over the last three years has averaged between US$8–US$10 billion, with a portion earmarked specifically for oil installation security. In addition, an average of US$1.2–1.5 billion is spent annually on the various security-related services provided by different ministries to safeguard the country's oil installations. The Saudi government takes the defense of its oil installations very seriously. At present, there is air surveillance of oil installations from helicopters as well as round-the-clock F15 patrols. Heavily equipped National Guard battalions stand guard on the perimeter. At any given time, there are between 25,000 and 30,000 troops protecting vital oil infrastructure. Saudi Arabia's oil terminals and platforms have its own specialized security units, comprising Saudi Aramco security forces, specialized units of the National Guard and the Ministry of Interior's petroleum installation security force. Additionally, the Coast Guard and components of the Navy protect the oil installations from the sea. Even if any attack occurred, the whole infrastructure would not necessarily collapse. However, there are some general vulnerabilities that need to be addressed, including the thousands of kilometers of pipeline system within the Kingdom.

Finally, the country must undertake other reform and development measures to ensure its overall economic security and stability. Notwithstanding its oil revenues, Saudi Arabia, on account of its centrality in the Middle East and its importance to the global economy, must ultimately embrace diversification as the only way to sustain the high oil prices. Without suitable diversification Saudi Arabia would not be able to fuel the huge economic boom and high growth required to meet domestic needs. Therefore, certain social reforms are at least as important as political reforms and despite the many positive steps that the Kingdom has taken over the last three years, much remains to be done.

US Dependence on Gulf Oil and Gas

Saudi Arabia's pivotal role as a global energy player and swing producer is further enhanced by its importance as a major US supplier. Aloulou Fawzi sheds light on the share of the Gulf in future US energy requirements. According to the Energy Information Administration's (EIA) annual outlook for energy markets through 2025, the US oil import dependence is forecast to grow from 56 percent to 68 percent in the reference case. The Gulf's market share in US oil imports is forecast to grow from 20.4 percent in 2003 to 30 percent in 2025.

The future growth in US natural gas supplies is also projected to depend on more imports from the Arabian Gulf, mainly liquefied natural gas (LNG). About 50 percent of the US natural gas imports are projected to take the form of LNG by 2010 and 40 percent of this could come from the Gulf region.

In the United States, the total energy use through 2025 will grow at an average annual rate of 1.4 percent, slightly less than half the projected rate of economic growth estimated at 3.1 percent. Fossil fuels – oil, coal and natural gas – are expected to account for an increasing share of total energy use. Petroleum demand is driven by growing transportation uses while natural gas demand is stimulated by the power generation and industrial sectors. Additions to gas-fired power generation capacity in the United States have outstripped electricity demand growth over the last 5 years and the EIA expects that the utilization of this capacity will rise as electricity demand continues to grow.

US coal consumption is projected to increase by 1.5 percent per year. Currently, about 90 percent of coal is used for electricity generation. The increased coal use reflects higher utilization rates at existing plants as well as the addition of new plants. US nuclear energy use is also expected to grow but at a very slow rate.

US domestic natural gas production is expected to rise more gradually than consumption does over the forecast period, rising from 19.1 trillion cubic feet in 2003 to 21.8 trillion cubic feet in 2025. The shortfall between consumption and production levels is made up by increasing reliance on natural gas imports.

Total US natural gas demand is expected to increase from 21.9 to 30.7 trillion cubic feet between 2003 and 2025. Growth in demand for electricity generation and industrial applications accounts for about 75 percent of the projected growth in total natural gas demand. Meanwhile, total US natural gas supply is projected to increase at an average annual rate of 1.4 percent per year between 2003 and 2025, from 22.4 to 30.6 trillion cubic feet.

Imports are expected to be priced competitively with domestic sources of natural gas and LNG is expected to account for most of the projected increase in net imports. By the end of the forecast, sufficient new LNG terminal capacity comes online to allow net LNG imports to increase from 440 billion cubic feet (bcf) in 2003 to 6.4 trillion cubic feet in 2025. By 2015, net LNG imports are expected to equal 15 percent of total US gas consumption, compared to 2 percent in 2003. Net LNG imports are expected to rise from 13 percent of net imports in 2003 to 62 percent in 2025.

Based on signed contractual commitments, Qatar will be producing 68.9mt per year of LNG by 2010, positioning itself as the world's leading LNG supplier. About 23.4mt per year is intended for the US market through ExxonMobil and ConocoPhillips. The US markets will import 2.5 trillion cubic feet of LNG or the equivalent of 52 million metric tons per year of LNG in 2010. More than 40 percent of US LNG needs could come from the Gulf region.

Gulf Supply and Asia-Pacific Oil Demand

On the supply side, the Gulf's role in meeting the growing energy import needs of the Asia-Pacific region is assuming ever greater significance. This is one of the world's fastest growing regions both in economic terms and in energy consumption. However, it is deficient in oil and gas resources, and thus heavily dependent on imported energy. Kang Wu and Jit Yang Lim stress that the Asia-Pacific is already the world's largest energy-importing region and its import levels will grow steadily over the next ten years and beyond.

Overall, the Asia-Pacific region faces a huge discrepancy between its primary fossil energy consumption and production. In 2004, the Asia-Pacific region accounted for 29 percent of the world oil consumption, yet its share of world oil production was only 10 percent, leading to high oil levels of import dependence. Even in natural gas, the Asia-Pacific region's global consumption outstripped its production in 2004, but the gap was much narrower than oil.

The Asia-Pacific region accounts for two thirds of the global trade in liquefied natural gas (LNG). Japan is the world's largest importer, followed by South Korea, Taiwan and India, while China is on course to become the fifth LNG importer in the region. LNG imports to these countries come largely from other Asian countries. However, Asian importers are acquiring more LNG from the Gulf and the Russian Far East. Over the next ten to fifteen years, FACTS projections indicate that LNG imports to Asia will continue to rise faster than the growth of oil.

The Asia-Pacific region has generally served as the engine of growth for global oil demand. Over the last 15 years, the Big Four Asian consumers – China, Japan, India, and Korea – have accounted for a little over 70 percent of total oil demand in the region. This is expected to remain unchanged in the future, through 2015. Sustained high oil prices are likely to slow down the oil demand for some Asian countries, especially those with a deregulated market, which has left consumers exposed to the relatively high oil prices. In the longer term, as the Asia-Pacific economies become more developed, oil demand outlook is not likely to grow as spectacularly as in the early 1990s. However, the region will still play a major part and account for about half of the world demand growth in the coming decade.

With rapidly growing oil demand and flat oil production, the Asia-Pacific region has a high ratio of oil import dependence. Of all the oil supply sources, the Gulf remains the most dominant. The Asia-Pacific region is heavily dependent on Middle Eastern crude and in 2004, almost three quarters of its 15.4 million bpd import requirements came from the Gulf. The Asia-Pacific and the Middle East are not only closely intertwined in terms of their crude trade, but also in their petroleum product markets.

China remains a major driving force behind the Asia's higher oil requirements and deepening import dependence on the Gulf. In 2004, the Gulf accounted for 45 percent of China's crude oil imports and this share is projected to reach 61 percent in 2010 as overall crude imports increase. If refined products are included, the Gulf's share in China's overall oil imports will be even higher. With its huge additional oil import requirements, China's rising dependence on the Gulf region constitutes one of the most important developments in Asia for the next decade and beyond.

Gulf Strategies for Ensuring Demand Security

In the energy scenario of the 21st century, how will the Gulf region fare with respect to demand security? Much depends on the relative global importance gained by different energy sources during the course of the century. Based on environmental and economic considerations, the long-term future for oil will be squeezed between the solidly rising annual coal supply over the 21st century and the dynamics of the global gas industry with an expected increase in its annual supply. In Peter R. Odell's analysis, the future of oil will essentially be limited by demand considerations, including geopolitical ones, and potential supply-side limitations will present only a secondary issue.

Natural gas will likely overtake coal as the second most important global energy source by the mid-2020s. This development reflects the near-quadrupling of worldwide proven gas reserves and the rapid expansion of European and other gas markets between 1975 and 2005. The industry now has a generous reserves-to-production ratio of 65 years. Thus, gas will undoubtedly be the fuel of the 21st century although production in the early decades is likely to be limited by demand, rather than by resources. The Gulf region, with almost 40 percent of the world's current proven reserves of conventional gas, should be able to greatly enhance its current contribution (of just 10 percent) to global gas supply.

Meanwhile, recent political events and extraordinary market developments have accentuated long-held western fears for the security of oil supplies from the Arabian Gulf region. Policy makers in many countries are already actively seeking to minimize dependence on oil, particularly Gulf oil. Present and potential supplies of both conventional and non-conventional oil in non-Gulf locations could thus be preferentially exploited, despite much higher costs of production and/or transportation. Additionally, greater efforts are being made to speed up the substitution of oil by natural gas, which is geographically more dispersed and environmentally more friendly. In this context, the Arabian Gulf's gas reserves need to be more effectively and extensively exploited through pipelines and maritime links with European and Asian markets.

Gulf interests in the long-term exploitation of its large oil resources are in the process of being undermined by the recent dramatic expansion of electronic crude oil market exchanges (particularly the NYMEX and the IPE). Trading in "paper" barrels on these exchanges, principally by "non-commercial" participants, is fundamentally undermining producer–consumer relationships and is making oil pricing an almost entirely speculative phenomenon. Gulf oil producers need to restrain this in their long-term interests, by returning exclusively to direct relationships with "physical" oil purchasers to avoid losing economic rent to the speculators and being able to stabilize prices at levels acceptable to both exporters and importers, thus securing oil market expansion.

Such an approach to crude oil pricing and marketing should become increasingly likely as international oil trade becomes state-oriented, rather than private sector-oriented. Concurrently, national oil corporations in the Gulf producing countries should also increasingly accept joint-venture investments from those developing countries, notably China and India, in which most incremental growth in global oil demand will occur over the coming decades.

KEYNOTE ADDRESS

Gulf Oil and Gas: Ensuring Economic Security

H.E. Sheikh Ahmad Fahad Al Ahmad Al Sabah

We have gathered here at time when the issue of achieving economic security for the oil producing nations and the world at large has become a matter of great concern. World oil markets have rarely witnessed such turbulent times as in the last two years. Oil prices have not only hit record levels repeatedly, at least in nominal terms, but there is also a lack of clear consensus on just what levels they will reach in the future.

As with previous oil shocks, hasty and often over-simplified accusations are directed at the oil producing countries, especially OPEC Members, as being mainly responsible for the current rise in oil prices. I would like to take this opportunity to set the record straight. Unlike the previous oil shocks, current high oil prices are the outcome of the recovery of the world economy, which has led to high oil consumption and demand. As you are all aware, such high demand has taken everybody by surprise, including the most experienced and informed oil market experts.

This demand increase has meant a significant reduction in global spare production capacity, which further feeds oil price volatility. Although the world economy has exhibited, at least until now, a greater degree of resilience than before in withstanding higher prices, one should not underestimate the harmful effects of a volatile oil market on the security of the world economy at large, including the economies of the oil producing and exporting nations. We in OPEC, contrary to widely held belief, understand fully these effects and are aware that they hurt both the consuming and the producing nations. Of course, there are some who are at times envious of the producing nations, which they regard as being the sole beneficiaries of high oil prices and therefore responsible for orchestrating them. It is not surprising that reports often view OPEC as being responsible every time there is a rise in oil prices. In this context, frequent calls are made for OPEC countries to urgently increase their production. This time, the situation is slightly different, in the sense that although accusations against OPEC have receded, they have not disappeared altogether. In the last two weeks, such calls were repeated by some parties in the consuming countries. In fact, these calls came after months of what an organization official has described as an unprecedented "semi-disappearance" of OPEC from the headlines at a time of high oil prices.

Such disappearance is not because OPEC has lost interest in playing a vital and continuing role in stabilizing the oil market, nor is the organization becoming irrelevant to the market, as some like to state. Commentators have played down the OPEC role because it has become so apparent that our organization has done and continues to do all it can to restore market stability.

Allow me to emphasize a fact that seems to have escaped many analysts and policy-makers within the global oil industry, or has been overlooked, whether intentionally or otherwise, for a variety of reasons. High oil prices that harm the economies of the consuming nations do not serve the interest of the producing nations, especially OPEC members. As many past experiences have shown, such high prices will collapse sooner or later, for a number of reasons. It is therefore of vital interest and concern to OPEC countries to preserve a healthy world oil demand through an equitable price level that would not adversely affect such demand.

In recognition of this fact, OPEC has tried and continues to try to lower the current prices by every means available. It has increased its production ceiling on several occasions in the course of the current price rise and will raise the ceiling further whenever it thinks stability is needed in the market. In addition, OPEC members are responding positively to the current situation by producing at or near their full capacity in order to meet the expected demand. It is no secret that OPEC is currently producing over 30 million barrels per day (mbpd), which is 1.5 mbpd above the official ceiling. Apart from increasing production, OPEC is considering other means of bringing prices down to acceptable levels by assuring the market of its willingness to meet any demand.

Let me draw your attention to the fact that, notwithstanding the current prices, there has never been an actual oil supply shortage. Admittedly, this time there is a fear that demand will genuinely outstrip oil supply, which will result in shortages, but we can assure the market that we will spare no effort to restore stability to oil prices, not only in the short term, but also in the long term, through the expansion of production capacity. Of course, this will require substantial investment in the Gulf oil industry, and will entail greater cooperation between all the players in the international oil industry.

The Hurricane Katrina disaster in the southern United States and the OPEC response, which included a commitment to meet any supply shortfall, has demonstrated the organization's willingness to work with consumers. It has also shown that cooperation between producing and consuming nations remains vital in achieving a stable world economy.

---

* This Keynote Address was delivered on behalf of H.E. Sheikh Ahmad Fahad Al Ahmad Al Sabah, President of the OPEC Conference and Secretary General of OPEC (2005) and Minister of Energy of the State of Kuwait (2003–2006) by H.E. Sheikh Talal Khaled Al Ahmad Al Sabah, Managing Director for Petroleum Services, Kuwait Petroleum Corporation.

Section 1

OIL PRICES AND ECONOMIC SECURITY

1

Oil Prices and the Challenges for the Producers

Herman T. Franssen

The current high oil price environment and the anxieties surrounding oil supply security are in many ways reminiscent of the developments of the 1970s, when oil prices rose ten-fold following two major oil supply disruptions. However, there are also major differences between the events that occurred then and events occurring now.

The oil price hikes of the 1970s contributed significantly to the two worst post-World War II global economic recessions of the mid-1970s and early 1980s. It may be too early to judge the impact of the recent oil price hikes on the global economy and particularly on the least developed oil-importing countries but monetary and fiscal authorities in the major economies have so far succeeded in avoiding a major global economic downturn.

The oil price shocks of the 1970s were the direct consequence of supply disruptions caused by political events: the Arab–Israeli War of 1973–74 and the Iranian Revolution of 1979. Since the oil shocks were the result of regional politically-induced oil supply disruptions causing havoc to the global economy, the political fallout was understandably large. OECD countries vowed that this would never happen again and that action would be taken to reduce their dependence on oil, especially on imported oil.

The global economy proved that it could deal with the doubling of oil prices between 2003 and 2005 better than the supply disruptions of the 1970s, partly because the share of oil in the GDP of the OECD economies has fallen significantly since the late 1970s. The other major reason is related to the changing global capital–labor ratio, which has helped keep OECD inflation at low levels.

The oil price hikes since 2003 stemmed from unforeseen escalation in oil demand in 2004, unexpected crude oil and refining capacity constraints and a major oil and natural gas supply disruption in the United States (in the autumn of 2005) at a time when there was little global spare oil production and refining capacity.

In the aftermath of hurricanes Katrina and Rita, in the autumn of 2005, the impact of the refining capacity outages affecting oil, natural gas and natural gas liquids has been of the same magnitude as previous major oil supply disruptions. However, their lasting impact has so far proved less severe on the US and global economy than the supply disruptions of 1973–74.

The limited adverse impacts of the two devastating hurricanes on the US economy and the speed at which capacity was brought back on stream, is a credit to the ability of the US and global oil industry to manage major supply disruptions. While the International Energy Agency (IEA) responded by making crude oil and products available from emergency stocks, most of the product flow from Europe to the United States originated from the private sector and only a small percentage of the crude oil offered by the US Strategic Reserve was actually auctioned.

So far, Western government responses to the steep oil price rises in recent years have been muted, particularly when compared with the reactions following the oil price shocks of the 1970s. The subdued official reactions may be due partly to the fact that the high prices resulted from high oil demand and not from a supply shock. Moreover, the impact on the global economy has been modest due to the lower oil intensity of OECD economies and in contrast to the 1973–74 situation, OECD inflation and interest rates have been low. Oil as a percentage of GDP was 4.1 percent in the United States in 1974 but only 1.9 percent in 2004. In Japan, the decline was even sharper, from 4.4 percent in 1974 to 1.4 percent in 2004.

Figure 1.1
The Bulls Take Command

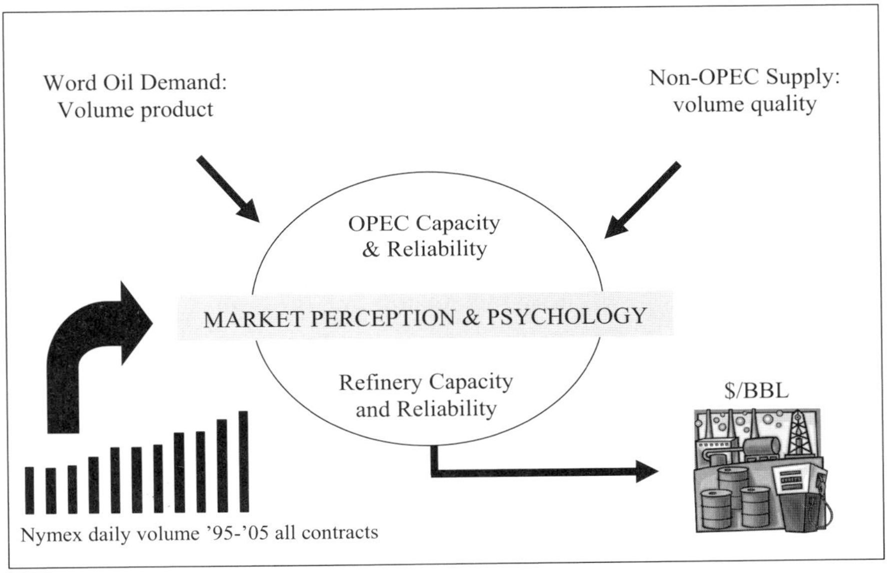

Source: Petroleum Economics Ltd, 2005.

The End of Cheap Oil?

For more than a quarter of a century after the Second World War, oil prices in nominal dollars were less than $3 a barrel. It is no exaggeration to state that the post-war economies of Western Europe and Japan recovered on an ocean of cheap oil, made possible by the vast discoveries of oil in super giant fields in the Middle East and North Africa by the Seven Sisters, the major international oil companies. Subsequent oil demand growth of 6–7 percent annually became unsustainable by the early 1970s and following the oil price shocks of 1973–74 and 1979–80, which contributed to two severe OECD economic recessions, major efficiency gains, fuel substitution and a huge increase in non-OPEC oil production, led to the collapse of the high oil prices in 1986.

Prior to the oil price hikes, the availability of cheap oil had encouraged its penetration in all sectors of the economy: power, industrial use, the residential/commercial sector and transportation. By 1973, when oil accounted for about half of OECD total primary energy consumption, energy industry experts had become aware that oil demand growth rates of 5–7 percent annually were unsustainable in the long run and at pre-1973 oil prices, new oil provinces in Alaska and the North Sea could not be brought on stream.

Following the tripling of oil prices in 1973–74, demand fell initially by more than 2.5 million bpd from 1974 to 1975 (a recession year) but continued its upward path in 1976, giving rise to the perception that there was a more or less fixed one to one relationship between economic growth and oil consumption. In reality, oil demand was price elastic but there was a time lag of several years before the impact of higher oil prices, combined with post-1974 OECD government policies to improve energy efficiency and fuel switching, became apparent. After the second oil shock of 1979–80, when oil prices once again tripled, the combined effect of the two oil shocks gradually began to take its toll on OECD oil consumption.

In the late 1970s energy industry experts were also bearish on future oil production growth. Most industry and government experts were of the opinion that despite high oil prices, global oil production was unlikely to increase much and that the world had no other choice but to quickly shift to other energy sources. A 1978 report by the UK Crown Agents for Overseas Governments and Administrations summed up the prevailing opinion as follows:

> Only in the last two years have most governments come reluctantly to accept that the days of cheap and continuous energy are over. Debate continues among the world experts as to exactly how much oil remains to be extracted and by what decade it will cease to be available; but all now are agreed that supplies of oil are running out fast and that we have to find alternative forms of energy if we wish life to continue as we know it.

A year later, a US government report entitled "The World Oil Market in the Years Ahead" referred to the predominant view among petroleum geologists that the chances of discovering enough quickly exploitable oil to offset declines in known fields, was very slim. The report concluded that if the Arabian Gulf countries and some non-OPEC producers continue to limit production (as was expected), world oil production would probably begin to decline by the mid 1980s.

A year after the US government report was published, BP issued a 1980 study entitled "Oil: Crisis Again," which basically conveyed the same pessimistic message. The BP report concluded:

> Within a decade the world is going to have to switch from supply, which grew at 5% per year over the past 50 years to 1973, to a supply, which will decline. The amount of oil available in the 1980s is not expected to greatly exceed that of 1978 because fewer large fields are discovered and decline rates are accelerating and by the 1990s oil production is actually expected to decrease.

When these studies were published, global oil consumption was about 65 million barrels per day (mbpd), some 20 mbpd below the 84 mbpd level witnessed in 2005. Global oil reserves were estimated at about 655 billion barrels in 1980. Global oil production between 1980 and 2003 amounts to 610 billion barrels and, even if we discount current global oil reserves by 100–200 billion barrels (taking into account overestimated "reserves" in some countries), global conventional oil reserves today would still amount to some 1000 billion barrels.

The high oil prices after the 1970s oil shocks led to major efficiency gains in all sectors of the energy economies in the OECD (and elsewhere) and major fuel substitution, particularly in the industrial and electrical power sectors. Substitution and efficiency gains cut OECD oil consumption by some 20 percent between 1979 and 1985 and the share of oil in OECD primary energy fell from 48 to 38 percent between 1973 and 1985. In the two decades following the oil shocks of the 1970s, average annual global oil consumption growth amounted to slightly more than one percent compared with the pre-1973 average of 6–7 percent.

Prior to the first oil shock of 1973–74, oil consumption was projected to reach 85 mbpd by 1985. The impact of two global economic recessions, slower OECD economic growth since the late 1970s, efficiency gains and fuel substitution, significantly slowed down oil demand growth. Instead of reaching 85 mbpd in 1985, global oil consumption will reach this level only in 2006 (21 years after it was supposed to have happened), due to the impact of much slower OECD economic growth after 1973–74 and higher oil prices leading to more efficient use of oil and fuel substitution.

On the supply side, loss of traditional low-cost production areas by the IOCs combined with high oil prices and the development of new technologies such as horizontal drilling and deepwater development and production technologies, led to an upsurge in non-OPEC exploration and production. Recovery rates of oil-in-place once thought to be limited to less than one third, have significantly improved across the board and loss of traditional upstream assets forced IOCs to look for oil elsewhere. In spite of the dire predictions about future oil production, global oil production today is some 20 mbpd higher than in the mid-1970s, and most of the net gains have been from non-OPEC oil producing countries.

Based on expert perceptions of low oil demand and supply elasticity, OPEC in the late 1970s and early 1980s decided to defend the price gains of the 1970s and began to cut production to maintain high prices. Lower OECD oil consumption in the first half of the 1980s and rapidly rising non-OPEC production forced OPEC to cut production. By 1985 about half of OPEC's oil production capacity was shut in and since Saudi Arabia had acted as the ultimate residual supplier among other residual suppliers, its production was down by two-thirds from the 1980 peak. In the summer of 1985, Saudi Arabia had great difficulty producing enough associated gas to power its electricity and water desalination plants. Sheikh Zaki Yamani viewed this as a wake-up call and convinced OPEC to shift its strategy of defending oil at any cost to regaining market share. Unless we understand the traumatic experience the almost (almost because other OPEC producers did shut in some production capacity as well) unilateral defense of high oil prices in the first half of the 1980s had on Saudi policy makers, we cannot fully understand their current skepticism towards renewed peak oil forecasts.

When OPEC, under Saudi leadership, decided to abandon its price support policy in favor of regaining market share, it still took almost two decades for OPEC to return to the 1979 oil production level and OPEC crude oil production capacity today (after 40 years) is still several mbpd below the level of the mid-1970s.

Oil Supply Security

Oil supply security became one of the issues dominating G-7 and OECD Ministerial meetings in the late 1970s. The prevailing perception among industry experts at that time was that oil prices would continue to rise and supplies would get scarcer. The global economy was expected to suffer from the inflationary impact of high oil prices as well as from the assumed contraction of global economic activity associated with the expected long term massive accumulation of foreign exchange in oil producing countries. The only way to counter this perceived threat to the global economy was for the US and other Western governments to develop energy strategies aimed at reducing oil imports and the use of oil altogether.

In the United States, Project Independence, completed in 1974, concluded that at $11 a barrel (in 1974 dollars) the country would be able to achieve energy independence by 1985. The Iranian Revolution of 1979, followed by the Iran–Iraq War of 1980–88 and the subsequent cut in oil exports from Iran and Iraq, further strengthened the arguments in favor of energy independence. Market responses to high prices supplemented by government policies initially caused OECD oil consumption to fall by some 20 percent or 7 mbpd in the first half of the 1980s and oil imports fell sharply in the United States (from 7.8 to 4.7 mbpd between 1979 and 1985) and Western Europe (from 12.6 to 8.2 mbpd over the same period).

Even import-dependent Japan was able to cut oil imports by 1 mbpd over this period.

By the year 1985, the demand for OPEC oil had dropped to half the 1979 level and after OPEC decided at its December 1985 conference to stop the erosion and increase market share instead, oil prices collapsed by the summer of 1986. OPEC struggled through much of the 1980s and 1990s to manage its surplus capacity in such a way as to stimulate demand by maintaining a price target sufficient to support the minimum budgetary requirements of the members, without encouraging consumers to shift away from oil and IOCs to develop high cost alternatives to oil. Between the mid-1980s and the early part of this decade, markets were adequately supplied with cheap oil in a range of $15 to $20 a barrel and core OPEC states (GCC members of OPEC) maintained adequate spare capacity to cope with unexpected supply disruptions.

After years of adequate oil supplies at moderate prices, the issue of oil supply security almost disappeared from the radar screen. During the Cold War, the West had always feared the possibility of a Soviet intervention with Middle East oil supplies. When the Soviet Empire collapsed in 1989, the residual fear of oil supply disruptions disappeared and the IEA, the OECD's watchdog of oil supply security, focused increasingly on environmental issues.

Apathy about the future of oil demand, supply and prices and about supply security continued through the end of 2003. The year 2004 proved to be one of those unpredictable inflection points on the chart. Global oil demand rose by close to 3 million bpd, largely due to unprecedented consumption growth in China, other Asian countries and the United States, reducing global spare production capacity to less than two million bpd of heavy, high sulfur oil. Effectively there was no more spare capacity because there were no refineries left with spare capacity to turn the Saudi heavy, high sulfur crude into light, low sulfur products. Market fundamentals pushed prices up to the highest level (in real terms) since the early 1980s.

Figure 1.2
World Refining Capacity vs. Product Demand

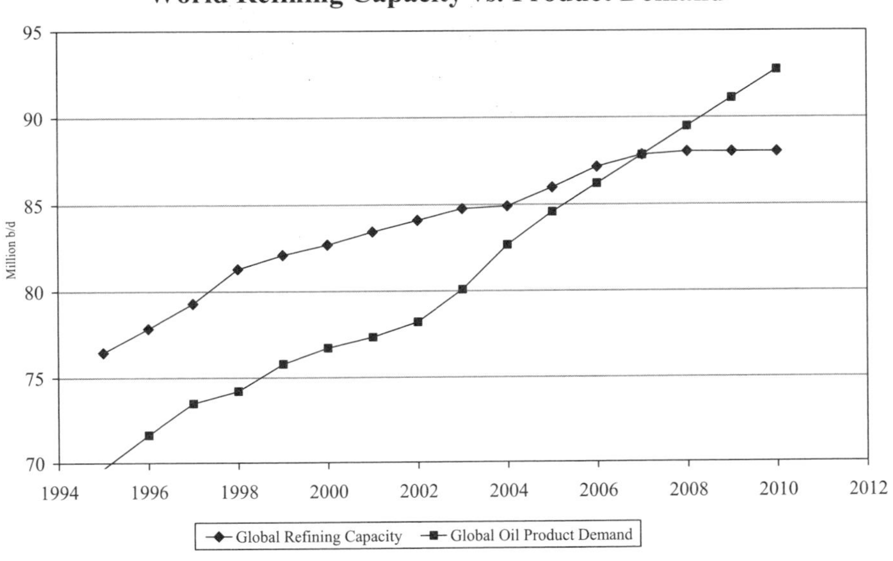

Source: Petroleum Economics Ltd., 2005.

The End of Cheap Oil Revisited

Medium and long term economic models have never been very good at predicting inflection points. Few predicted the major shift in economic growth in the OECD, from the high, pre-1973 growth rates to the low-to-modest post-1973 economic growth rates. Throughout the 1980s until the year 1989, it was generally assumed that Japan's economy would grow faster than any other region in the OECD. When Japan's "bubble economy" burst in 1989, signaling a prolonged period of slow economic growth, it came as a great surprise to most international economists. Finally, it was generally accepted in the 1980s that the EU economy would soon overshadow US economic growth, which at that time was suffering from a major decline in its industrial "rust belt." When the US economy recovered in the 1990s and showed unparalleled expansion in the second half of the 1990s, many economists again assumed that the

United States was bound to continue to grow much faster than other developed regions in the world for years to come.

Following years of a predictable steady growth pattern, inflection points have once again proved to be very difficult to predict in regional and global economic activity as well as other areas such as changes in the global oil market. Until the late autumn of 2003, for example, the consensus among oil experts was that over the next five years and perhaps until the end of the decade, oil prices would likely remain in the low twenty dollar range, with a small change of a price collapse in 2005 or 2006. These oil price forecasts were based on the perception that global oil consumption growth would remain low, non-OPEC production growth would be high, OPEC oil production capacity would rise steadily and Iraq would stage an early return as a major oil producer after the middle of this decade.

Oil market developments in 2004 came as a complete shock to most oil analysts. With perfect hindsight, it was easy to explain why oil prices rose by an average of about $10 a barrel in 2004. China's phenomenal economic growth and continued buoyant economic growth in the rest of Asia and the United States, sharply increased global oil consumption, reduced global spare capacity to a trickle and whatever was left could not be refined due to shortages of upgrading capacity in refineries. Few, however, had even remotely thought that such a sequence of developments in the oil market could realistically occur and when it did, the perception was that it could not continue for long.

The consensus view that oil price would remain high was formed when, despite a significant reduction in demand growth in 2005, oil prices rose on average by another $15 a barrel. Most energy economists still had difficulty in accepting that high oil prices would persist and by the autumn of 2005 most of them projected that prices would fall to the $40-plus level in 2006. By early 2006, however, with oil prices remaining high despite worsening oil market fundamentals, the consensus shifted to mid-term (next 5 years) oil prices in $45–$55 a barrel range.

The End of Cheap Oil or Return to Price Cycles?

The pre-2004 industry consensus was that oil prices would hover in the low twenty dollar range at least through 2008. It was based on perceived market fundamentals and on the assumption that at that time almost all conventional oil could be profitably developed at around $20 a barrel.

While the newly emerged industry consensus calls for $50-plus oil for the next few years, analysts do not necessarily agree on the reasons for the higher medium term price forecasts. Nor do they agree on the sustainability of the current high price environment. The oil industry may have come to accept the prospect of $50-plus a barrel for West Texas Intermediate (WTI) for the medium term but the hurdle rate, the price level used by upstream companies to determine investment decisions, is closer to $25 a barrel. Hence, projects with overall costs significantly in excess of $25–$30 a barrel are unlikely to be financed unless they have a quick payback period.

Those who argue that the era of "cheap oil" (not clearly defined but usually assumed to be less than $20–$25 a barrel) is over and oil prices will remain above $40 a barrel, give some or all of the following reasons: peak oil, emerging demand from Asia, particularly China and India, less room for fuel switching, reduced IOC access to oil resources, escalating exploration and production (E&P) costs, resource nationalism, OPEC determination to set a new floor under oil prices in excess of $40 a barrel, lack of spare capacity, shortages of upgrading capacity in refineries and finally, internal political and geopolitical issues.

Those who argue that the oil market is cyclical and that high prices will slow down demand growth, increase production, promote long term efficiency practices and shifts to less expensive fuels, believe that prices will eventually fall considerably below the current level. They point to past oil price cycles to make their point. High prices have stimulated supply innovation and constrain demand growth. Few oil analysts believe that prices will return to the pre-2003 level of $25 a barrel or less on a sustained basis but most agree also that high prices will stimulate efforts leading to more efficient use of energy and to the development of technologies to develop substitutes to conventional low-cost oil. In the first few years following the oil price increases of 1973–74, global oil demand kept rising. It was not until after the second oil shock in 1979 that efficiency gains, fuel substitution and higher non-OPEC production began to have a major impact on global oil consumption.

Hence, both views may turn out to be correct but at different times. Global oil demand growth in 2005 fell to less than half of the unusually high level of apparent oil consumption in 2004. Demand was down across the board in all parts of the world but the most striking was the complete reversal in Chinese oil consumption. Chinese data suggest that in 2005 their economy grew at close to 10 percent on an annual basis while oil consumption almost remained flat. If true, this would be unprecedented in a developing industrialized economy. Most analysts believe the data is incorrect but all agree that Chinese oil consumption fell in 2005 partly because additional coal-fired power plants replaced diesel-generated power and there were restrictions on the availability of motor fuels, limiting transportation fuel growth. Despite lower Chinese oil demand, the market remained tight because non-OPEC oil production growth was very low in 2005 and demand for OPEC oil remained close to the group's production capacity.

Market fundamentals weakened during 2006 due to continued slower oil demand growth on the back of high prices and higher oil production. Non-OPEC incremental supply, (including NGL supply from OPEC) could turn out to be slightly higher than global oil demand growth, leaving demand for OPEC crude oil more or less unchanged from 2005, even though OPEC may add more than 0.5 mbpd of additional capacity. If oil production is not shut-in for geopolitical reasons in any one or more political hotspots in the world, the build-up of OECD commercial stocks could push stock levels to a five-year high during the summer. If that happens, OPEC may consider cutting back oil production later in the year to keep oil prices from falling.

Figure 1.3
Regional Oil Demand Growth (2000–2006)

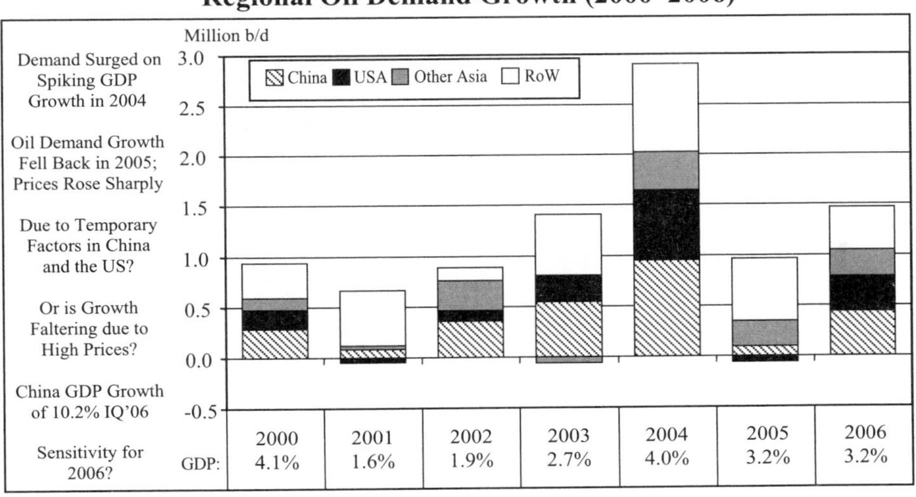

Source: Petroleum Economics Ltd., 2006.

The current high oil prices are only partially due to market fundamentals. The oil market in early 2006 is undoubtedly supported by a significant "fear premium" in the market, based on supply uncertainties from several countries, ranging from Venezuela to Nigeria, Iraq and Iran. If most or all of the possible threats to supply disruptions were removed, market fundamentals would put oil prices under downward pressure and OPEC would have to curtail production.

Oil market developments over the next few years to 2010 suggest the following trends:

- Modest oil demand growth averaging between 1.5 and 2 percent annually
- Significant growth in non-OPEC crude oil capacity, in particular from the Caspian region, West Africa, Brazil and Canada (oil sands) and continuing expansion of OPEC natural gas liquids (NGLs)
- Expansion of OPEC production capacity, particularly in Saudi Arabia and Nigeria but also in other member states such as Algeria, Libya

and the smaller GCC producers (Iraq has considerable potential but due to the current political situation, the investment climate is very uncertain).

Based on upstream projects under development, non-OPEC incremental oil (plus OPEC incremental NGL production), could be sufficient to meet most of the projected global oil consumption growth for the next few years. Since OPEC members are projected to add more than 2 mbpd of capacity over the next few years as well, spare production capacity may finally grow again to a more comfortable level. Spare capacity will revive OPEC's ability to better manage oil supplies and influence medium term oil prices.

Risk factors in the medium term include the possibility of slower regional or global economic growth or even a global recession, which could be triggered by a variety of events, ranging from the avian flu pandemic to heightened concerns about the unsustainable nature of a global economy dependent on only two engines of economic growth: China and the United States. The combination of a global economic slowdown, leading to lower incremental oil consumption and rising non-OPEC and OPEC production capacity, would lead to rising spare production capacity. Based on past experience, rising OPEC spare capacity, unless properly managed, could create significant downward price pressures on oil prices. Hence, there is a risk to OPEC of a downward oil price correction in the medium term in case of a global economic slowdown. Moreover, maintaining oil price stability in the new price range ($40–$50 a barrel) currently under discussion by OPEC may once again require the group's active supply management over the next few years.

Potentially offsetting the impact of non-OPEC and OPEC oil production capacity growth are the following factors:

- Higher oil demand growth caused by continued strong global economic growth, spearheaded by a multiple Asian economic steamroller (China and India), higher growth in Japan and the European Union and a less disruptive solution to the US dual deficit

- Significant delays in non-OPEC oil production capacity growth for technical or other reasons
- Higher oil depletion in producing fields. Most analysts assume between 3–5 percent depletion in producing fields but the Chief Executive Officer (CEO) of Schlumberger last year at the Oil & Money conference in London said that it could be as high as 8 percent. The difference between 3 or 8 percent global depletion is equal to 4 million bpd or the size of all of Iran's oil production.
- OPEC determination to maintain high ($50-plus) oil prices for internal economic reasons
- Supply disruption and/or slower capacity growth in OPEC due to geopolitical constraints
- The combination of sustained high oil demand growth and slower non-OPEC and OPEC capacity growth as well as production constraints for geopolitical reasons, could keep oil markets tighter and prices on the high range of current estimates.

Oil Market Developments in the Post-2010 Period

Current perceptions of oil market trends in the post-2010 period are as follows:

- Rising demand in the developing world, spearheaded by demand growth in China. Eighty percent of the world's population currently consumes on average less than 2.5 barrels per person per year, or less than one-fifth of the average oil consumption in the OECD. With economic growth oil demand is expected to grow for decades to come. The next decade may require major innovation to improve energy significantly and in particular, oil utilization technologies to cope with higher prices and the impact of global warming.

- By the early part of the next decade, more oil producing countries are projected to reach a production plateau for technical and other reasons and some high reserve OPEC producers may decide on an oil production conservation policy, limiting production considerably below the technically feasible production level. New cost-cutting technologies may be required to develop large volumes of non-conventional liquids for the transportation sector.
- While prospects for Arctic and ultra deepwater oil and gas as well as non-conventional liquid hydrocarbon developments are promising, they may not come on stream fast enough to offset a possible plateau in conventional oil production.

Figure 1.4
Estimated Oil Demand Per Capita (2025)

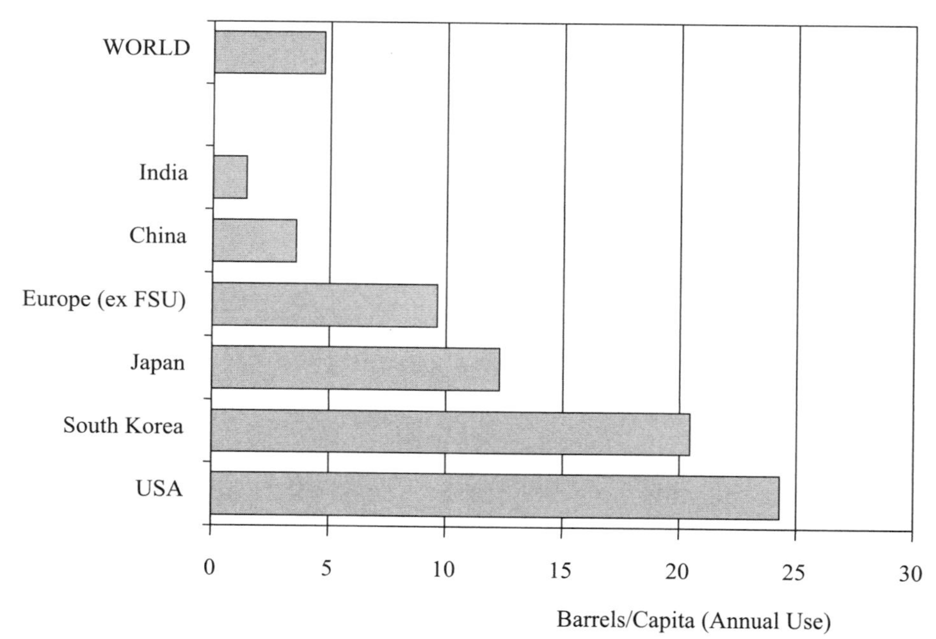

Source: Petroleum Economics Ltd., 2005.

Rising Oil Demand

The emergence of China, and to a lesser extent, India, has added a new dimension to the oil market. In the 1970s and 1980s, economic growth in the Asian "tiger economies" caused developing Asian oil demand (excluding China) to grow from 1.7 mbpd in 1973 to 6.5 mbpd in two decades despite higher oil prices. Today, China and India are the new "tiger economies" with a combined population of more than 20 times the population of the earlier tiger economies. The impact of high economic growth in these two countries on energy and oil consumption growth is going to be very significant.

Figure 1.5
World Oil Demand

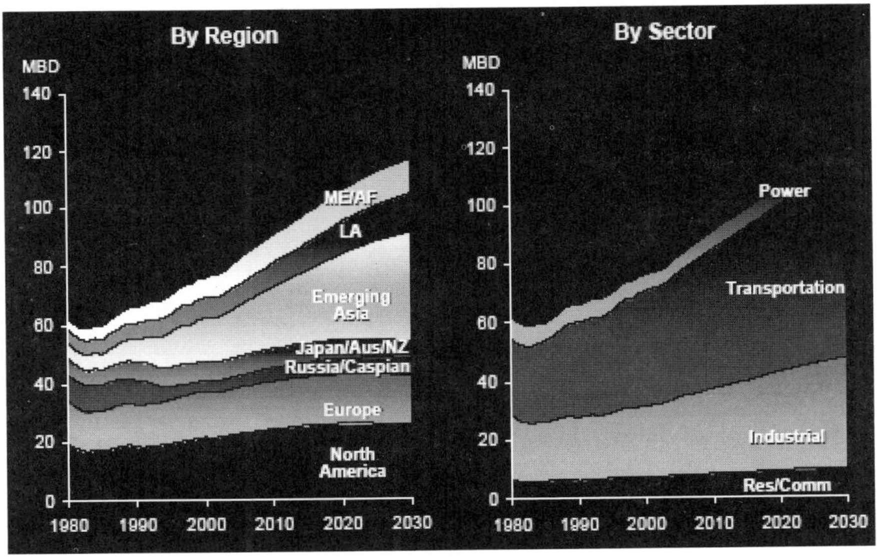

Source: ExxonMobil, 2005.

China's oil consumption rose from 1mbpd to 3 mbpd over that same twenty years. Continued strong economic growth has more than doubled oil consumption in China since the early 1990s. In addition to the industrial use of oil, the transportation sector is undergoing rapid growth.

Annual car sales in China is expected to rise from about 2 million in 2000 to 10 million by 2010 and over a period of fifteen years, the Chinese vehicle stock could increase from about 30 million in 2005 to over 120 million by 2020 if China follows a similar pattern of transportation policy as its Asian neighbours.

Figure 1.6
Projected Chinese Vehicle Stock (2000–2020)

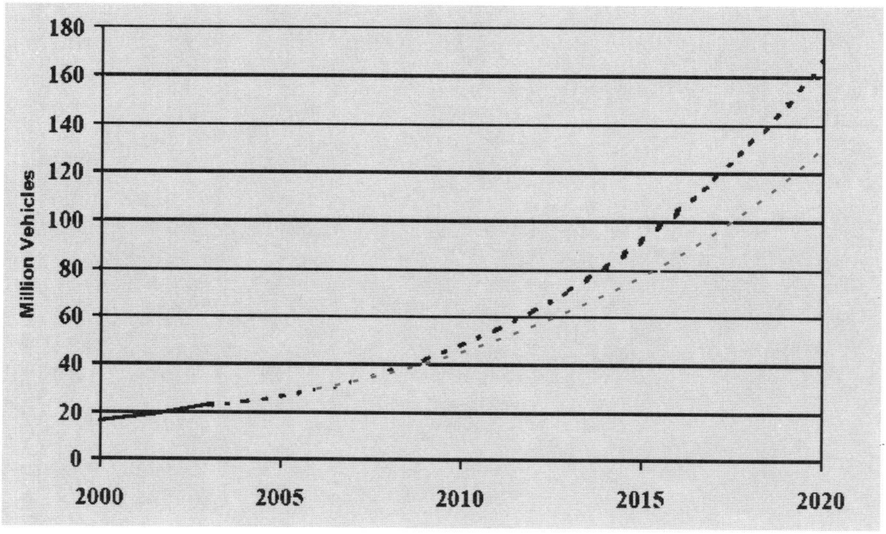

Source: US Government, unpublished document, 2004.

Other Asian economies are not growing as fast as China. However, there is little doubt that Asian oil demand will continue to grow in line with economic development and may double to over 30 mbpd by 2025.

Oil consumption in other developing regions such as Latin America and in particular the Middle East is also projected to show strong growth. By contrast, US oil demand growth may slow down from previous years and the European Union and Japan may show very little growth in future oil consumption due to a combination of high consumer taxes and government policies aimed at promoting the reduction of oil consumption through energy efficiency and the production of alternative transportation fuels.

Long Term Oil Production Trends

In recent months, Chevron has taken out full-page advertisements informing American consumers that in that past two decades the world has been consuming double the quantity of oil that has been discovered. While petroleum geologists and engineers in the oil industry are divided on when non-OPEC oil production is likely to reach a plateau and how much more production capacity to expect from OPEC, there is general agreement that the era of easy access to cheap oil is over. Most industry forecasts now show non-OPEC crude oil production to peak in the early part of the next decade and if OPEC NGLs, condensate and non-conventional oil are included, the plateau is projected for sometime by the middle of the next decade.

Figure 1.7

Past Discovery and USGS Estimates of Future Discoveries

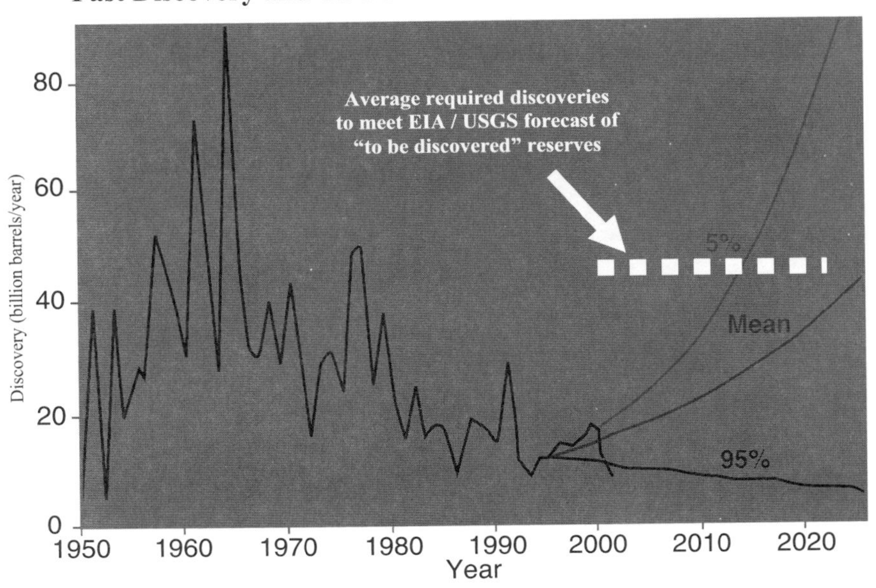

Source: Sadad Al Husseini, Oil & Money conference presentation, London, October 2005.

Industry experts range from those who believe that global conventional oil production (there are differences in the definition of conventional oil) is nearly peaking to those who believe that a conventional oil peak is still several decades away. Much depends on the perception of decline rates in producing fields (expert opinions vary from 3–8 percent average decline in producing fields) and the volumes of undiscovered recoverable oil left in the world.

Figure 1.8
The Oil Supply Challenge

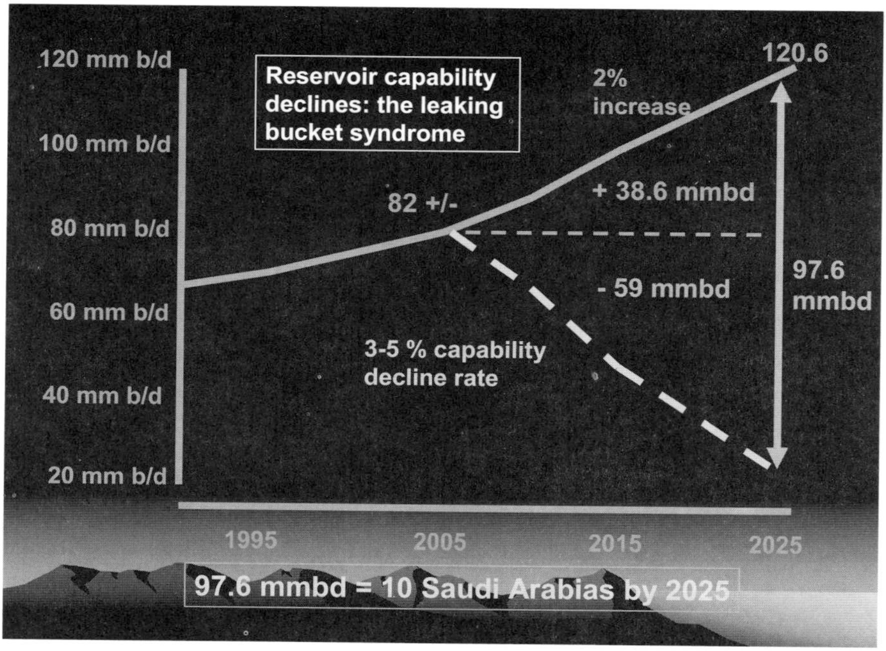

Source: Sadad Al Husseini, Oil & Money conference presentation, London, October 2005.

While there is no definitive answer to these questions, suffice to say that the task that lies ahead is truly immense. If global oil consumption were to grow at 2% annually to about 120 mbpd by 2025 and global oil depletion is between 3 and 5 percent annually, the global oil industry would have to find enough new oil to produce close to 100 mbpd of new oil by 2025. The challenge, if it can be met is to find the equivalent of 10 Saudi Arabias (in terms of production capacity) over the next quarter of a century. It is not inconceivable that global conventional oil production will peak at a much lower level, requiring massive new efforts to significantly improve efficiency in the transportation sector and development of new sources of fuel, ranging from gas-to-liquids (GTL) and coal-to-liquids (CTL) to biofuels.

The Geopolitics of Oil

One of the major challenges for the oil industry is to secure the flow of oil supplies from major oil producing regions of the world to end use markets. In recent years, oil supplies have been disrupted in Venezuela, Nigeria and Iraq and current concerns about supply security have added a premium to the oil price. Many major oil producing countries are in the midst of major internal socio-economic and political changes, which have from time to time led to supply disruptions. Such disruptions proved manageable as long as OPEC had adequate spare production capacity (as evidenced during the Iran–Iraq War, the Kuwait invasion of 1990, the Venezuelan strike and the Iraq invasion of 2003). However, supply disruption has and will lead to major oil price spikes when there is little or no spare oil production capacity in the global supply system. The inability of the global refining system to crack and treat spare capacity of heavy, high sulfur crude oil has further complicated the issue.

Fear for the possible impact of destabilizing internal developments as well as continued regional conflicts and friction between some major oil producing countries in the Middle East and major oil consuming countries, will continue to influence oil psychology and in some cases the level of actual production.

If non-OPEC conventional oil production were to peak sometime after the turn of the decade and OPEC's capacity growth were to slow down, either for technical reasons or for the purpose of conserving scarce resources, the pressure for access to oil may lead to conflicts between oil consuming countries and regions. Competition for access to oil between traditional and new buyers of oil from large oil producing regions has already begun and is likely to intensify in the next decade. Traditional alliances may weaken and be substituted by new alliances between major oil producing and consuming countries and regions. New, emerging economies may develop blue water navies to protect vital sea lanes from the oil producing regions to their shores.

Figure 1.9
World Liquids Production Outlook

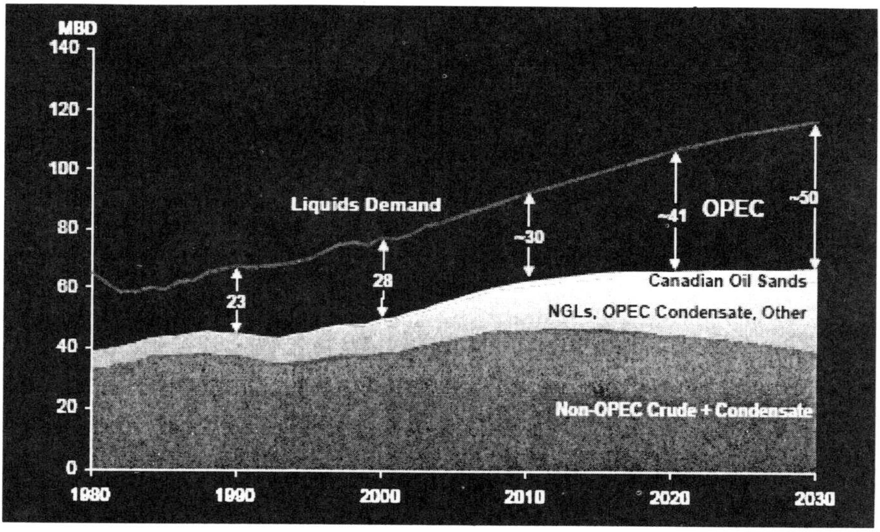

Source: Exxon Mobil, 2005.

The world may be entering a new era of slow but gradual transition from old to new sources of fuels. Today, the average oil consumption of the 5.5 billion people living outside the OECD, is approximately 2.6 barrels per person per year. The average oil consumption in the OECD is 16.5 barrels per person per year. To double per capita oil use in the developing world to about 5 barrels per person per year (about the current level of per capita oil consumption of Thailand and only one-third of the current average per capita oil consumption in the OECD) by 2020 or 2025, would require about 90 million bpd. Assuming there would be no change in OECD oil consumption of 50 mbpd, the global demand for oil could reach 140 mbpd.

Current estimates of global production of conventional liquids (oil and NGLs) and liquids from tar sands, extra heavy Orinoco oil and GTL may reach a plateau somewhere between 100 million bpd (the pessimistic view) and 120 million bpd (the optimistic view). However, even in the high case, global oil production would fall short of potential demand growth by 2020–2025.

In a September 2004 IMF report, Dr. Raghuram Rajan, Director of the IMF Research Department wrote:

> We have to realize that in the long run, and without dramatic technological change, it will simply be unsustainable for every Chinese or Indian household to consume as much energy and as inefficiently as the average American suburban household. Clearly, all countries have to increase incentives for both conservation and energy efficiency, while removing unnecessary impediments to exploration and production.

The United States consumes about 26 barrels of oil per person per year, the EU and Japan use about 15 barrels per person per year while the OECD countries average about 18 barrels per person per year. The world today consumes about 85 million bpd. If global per capita oil consumption were to mirror US per capita consumption, global demand for oil would be in excess of 300 million bpd, almost three times the expected global oil production plateau of 100 to 120 million bpd.

Hence, for the world to reach the US, EU or Japanese levels of oil consumption at any time in the future using current technology is simply out of the question. Even at Thailand's per capita level of oil consumption of about 4.5 barrels person per year, the world would require a level of oil production of 140 million bpd, which is clearly out of sight. In the decades to come, the global economy will move hundreds of millions of people into a global middle class (India and China's middle class is currently estimated at close to 500 million or 65 percent more than the entire US population). This middle class will be seeking a lifestyle similar to that of the average Japanese or European urban dweller. To satisfy their requirements of electrical power and personal transportation will be a challenge of unprecedented proportion.

In summary, in the short term (up to 2006), the crude oil market seems well supplied as shown in the historical high level of OECD commercial stocks. The main bottlenecks are in the refining sector, where upgrading capacity, capable of turning high sulfur, heavy crude oil into light, low-sulfur products, has not yet kept up with demand. Concerns about oil supply disruptions in Africa and the Middle East in particular, has also contributed to keeping oil prices (WTI and Brent) around $70 a barrel. OPEC is producing all the crude oil the global refining system can absorb and several member states are adding export refining capacity of their own.

In the medium term, through the turn of the decade, additional non-OPEC and OPEC oil production capacity growth is expected to once again provide several million bpd of spare crude oil capacity. This would enable OPEC to manage supply in an effort to maintain a desirable oil price level (which has been unofficially defined on occasion as being around $50 a barrel for the OPEC Reference Basket of crude oils). There is some concern in OPEC circles that government policies in OECD countries to reduce oil imports from the Middle East will have an adverse impact on the demand for their oil at a time when they are making massive new upstream investments to meet projected future global oil demand. Hence, producers are as adamant about receiving some guarantees on "oil demand security" in the same way as consumer countries are pushing for oil supply security.

It may take several years more before global upgrading capacity in refining would have caught up with projected global oil demand growth and hence, product prices may continue to reflect the tightness in refining capacity. OPEC is making a major contribution to easing the tightness in the refining sector with firm plans to add massive additional refining capacity, in particular in the countries of the Arabian Gulf.

In the longer run, the demand for transportation fuels in the emerging economies of Asia, the Middle East and Latin America, will create a major challenge for OPEC, particularly for the high reserve, low consumption countries of the Arabian Gulf. On the demand side, there is little doubt that the middle class of the emerging industrializing, developing countries (the middle class in China and India together is larger than the entire North American population) desires modern standards of living, including the use of private cars. The pent-up long term demand for transportation fuels from the emerging developing countries is truly staggering and there is no substitution in sight for oil products to meet this demand. Non-conventional fuels (hydrocarbon and bio-fuels) will add supplies at the margin but oil will remain the main source of transportation fuel for decades to come.

The oil industry is divided on the ability of the world to expand conventional oil production and to increase non-conventional transportation fuels in a timely fashion to meet global fuel demand at "reasonable" prices. The current prevailing view is that non-OPEC oil production, which has been growing rapidly for decades, may reach a plateau (some argue that they will peak) around the turn of the decade. If that were to happen, incremental demand for oil will have to be met increasingly from OPEC, and in particular, from Middle East OPEC sources.

The main challenge for the Arab Gulf states is to manage their oil reserves and resources in such a way as to meet global oil demand at "reasonable" prices, while preserving much of their resources for future generations. It may be noted that the countries of the Arabian Gulf in particular, have a limited ability to diversify their economies away from dependence on oil (and gas) income.

To enable future generations to benefit from oil (and gas) income, producers will have to settle for a long oil production plateau of several decades, based on a conservative assessment of sustainable production capacity. In view of the political pressures exerted by oil consuming countries from east and west, this will not be an easy task. In the end, however, even consuming countries will benefit from extended Arab Gulf production plateaus because it will allow consuming countries to gradually add non-oil fuels rather than be forced into massive crash programs if Arab Gulf oil producers were to push production to the technical limit, followed by short plateaus and subsequent production declines.

2

Gulf Economic Security in the Age of Globalization*

Tarik M. Yousef

By the end of 2005, the Gulf economies had registered the third consecutive year of strong economic performance.[1] The exceptional conditions of the past few years have been reflected in rising incomes, surging fiscal and external surpluses, shrinking public debts and rising levels of foreign reserves. Already in 2004, the aggregate nominal gross domestic product (GDP) for the Gulf Cooperation Council (GCC) countries rose by 18 percent to US$470 billion, and GDP per capita averaged about US$14,000 for the region as a whole. The GCC's aggregate fiscal surplus reached close to US$64 billion (14 percent of aggregate GDP), and the aggregate current account balance increased by more than 60 percent to US$81 billion (17 percent of GDP).[2] Estimates for 2005 and projections for 2006 suggest continued strong performance with aggregate real GDP growth of around 6 percent.

At the core of this boom is the combination of rising oil revenues, expansionary fiscal policies and massive liquidity flowing through financial markets (See Figure 2.1). So long as oil prices remain at their present high levels, all indications suggest that the Gulf economies will sustain their recently found prosperity for a while. Yet herein lies a danger as well.

* This paper draws on and updates research completed in 2002–2004 while the author was a Visiting Professor in the Office of the Chief Economist in the Middle East and North Africa Region at the World Bank. He is grateful to Dipak Dasgupta, Nadereh Chamlou and Mustapha Nabli for helpful discussions and to David DeBartolo for excellent research assistance.

What if oil prices take an unexpected downward plunge, as they have habitually done in the past three decades? More worrisome, what if the current wave of prosperity postpones the adoption of the structural changes needed to prepare the Gulf countries for the challenges they face in the globalizing world economy? After all, oil booms have traditionally provided breathing space for governments and delayed the implementation of reform programs aimed at securing long-term prosperity.[3]

Figure 2.1
GCC Oil Revenues (1970-2004)

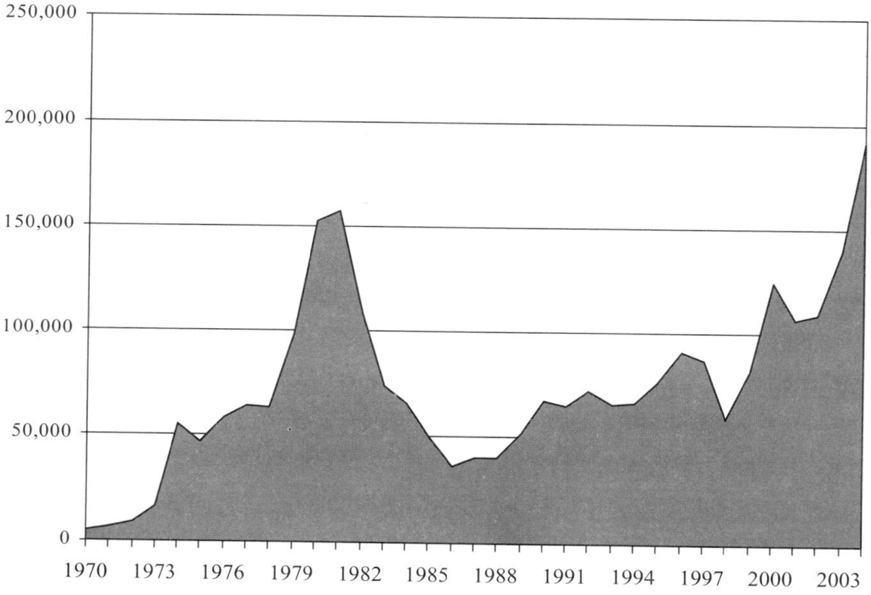

Source: World Bank, 2005

In other words, the exceptional economic conditions in the Gulf today are unlikely to last and may divert attention from important policy objectives that have eluded the region in recent history: long-term economic growth and stability. In this chapter, it will be argued that the pursuit of these objectives should guide any conceptualization of Gulf economic security in the 21st century. It examines not only the meaning of "economic security" for the Gulf countries in the era of globalization, but also whether globalization alone can address the Gulf's core development challenges like job creation and economic diversification.[4] It suggests a broad policy framework designed to address both these core development issues and globalization's unique challenges, thereby maximizing the opportunities and minimizing the risks of integration into the world economy.

Economic Security in the Gulf: An Elusive Target

The meaning of economic security has changed fundamentally over the past decade. Traditionally, economic security meant independence from external economic forces; it was a way of limiting a state's vulnerability to economic manipulation by rival states.[5] However, for the small Gulf economies, isolation and autarky were never viable options for achieving prosperity. These countries derive much of their income from exporting hydrocarbon resources to world markets and invest considerable sums of these proceeds in international portfolio and direct investment.[6] Moreover, with the importance of uninterrupted oil and gas supplies to the industrialized world, the Gulf countries have historically had little choice but to focus on the development of their hydrocarbon sectors.

Such interdependence with the external world tilted investment strategies in favor of the oil sector and constrained the possibilities for economic diversification without government intervention.[7] As a result, from the 1970s, several Gulf countries adopted development strategies based on state-led industrialization and state-supported private sectors. Subsequent decades witnessed significant advances in socio-economic indicators and rapid modernization. However, the long-term economic payoff from these strategies was inherently limited, on the one hand, by the volatility of oil revenues and, on the other, by the small size of domestic markets in most GCC countries. This became apparent in the mid-1980s when the sharp fall in oil prices led to a rapid deceleration in public expenditures, private investment and economic growth.[8]

Under these circumstances, economic security invariably required not only stabilizing oil prices but also expanding the size of the domestic market.[9] In the late 1980s, this implied pursuing projects aimed at regional economic integration through the GCC. The logic was straightforward: trade and investment integration would allow the Gulf countries to lower their dependence on oil revenues, minimize policy distortions and expand the size of the domestic market. Even in this situation, the long-term payoffs are not necessarily high. The similarity of economic structures and natural resource bases in the Gulf is believed to limit the potential gains from integration, especially with the exclusion of two large neighbors from this project, Iraq and Iran.[10]

This conceptualization of the policy dilemma facing the Gulf countries in their pursuit of economic security changed dramatically in the 1990s. The expansion of the world economy, made possible by multilateral trade liberalization and the advent of information technology, provided small states with the possibility of securing prosperity through global economic integration. The merits of globalization became all the more clear, given the strong growth record of the fast-integrating countries in East Asia and the stagnation of closed regions including in the Middle East. Thus, it appeared that economic security in the contemporary globalizing world economy required increasing openness and interdependence.

Economic Security in the Era of Globalization

Is globalization an adequate answer to future economic security needs in the Gulf? Recent discussions of economic security have stressed not only the long-term growth benefits but also the short-term risks posed by integration into the world economy.[11] Indeed, the 1990s have demonstrated the downside effects of globalization due to the volatility of commodity producers to unpredictable changes in demand, the vulnerability of capital flows to changes in market sentiment, and the unequal distribution of benefits and costs from trade integration. Thus, contemporary discussions of economic security in the era of globalization are concerned as much with managing these risks as they are with maximizing the benefits. The unique economic security challenges that globalization poses must be examined in depth, because they are directly applicable in the Gulf context.

In today's closely interconnected global economy, open national economies face significant economic security risks from shocks originating outside their borders. The Mexican and Asian financial crises of the 1990s are the most visible manifestations of this type of economic volatility that disrupted economies enjoying rapid growth. What made the shocks particularly worrisome was the spread of distress across national boundaries into neighboring countries and even regions far from their epicenter. These national crises sparked powerful global consequences, including in the Gulf region. For oil-exporting states, the sudden and unexpected decline in Asia's demand for oil in 1997–98 drove prices sharply downward, putting immediate stress on budgets, current accounts and exchange rates.

The flexibility of movement available to investors in developing countries is another aspect of this volatility. The rapid mobility, especially of short-term speculative capital flows – which is likely to intensify in the Gulf as countries further liberalize their investment regimes – makes countries vulnerable to the destructive effects of sudden reversals.[12] Moreover, as seen in the 1990s, the contagious effects of such reversals – as when events in one country lead investors to liquidate their holdings in other related or unrelated markets – can be harmful. The large correction to stock market valuations across the Gulf in March 2006 may have reflected precisely this mechanism. From a policy perspective, the difficulty of addressing such volatility through unilateral national measures has placed pressure on regional and multilateral institutions to address these transnational problems.

While the expanding world economy creates incentives for skill accumulation and provides opportunities for rapid job creation, the rewards anticipated from integration could be undermined by the volatility of international markets. Private sector employment, though more lucrative, is inherently less stable than public sector employment, which employs the bulk of nationals in the Gulf region. Moreover, the high potential for swift sectoral changes in a globalizing economy – away from traditional manufacturing and services, for example – jeopardizes the employment of those whose skills become obsolete. While educational systems can address the need for new skills in the long term, other policies are needed to address the short-term dislocations in labor markets.

The current phase of globalization is distinguished by the proliferation of non-state international networks, which can be either beneficial or detrimental to economic security.[13] On the one hand, the mobility of students, business people and civil society actors across borders can deepen a nation's understanding of the opportunities and constraints of globalization, but on the other hand, networks trafficking in drugs, humans or terror can have serious security consequences.[14] However, the measures that governments take to address national security vulnerabilities can themselves have detrimental effects on their economic security and that of others. For example, America's restrictions on student visas in the wake of September 11 or the opposition by members of the Congress to the Dubai Ports World's prospective acquisition of operations in several US ports have arguably done harm to the economic interests of all parties.

The contemporary era of globalization is also distinguished by the extent to which traditional ideas of state sovereignty are being replaced by shared international norms and practices. In the economic sphere, the "Washington Consensus" espoused by multilateral organizations like the International Monetary Fund (IMF), World Bank and World Trade Organization (WTO), is aimed at promoting long-term economic prosperity. Yet, the implementation of structural adjustment programs in several Middle Eastern countries in the 1980s and 1990s was greeted with protests, demonstrating the degree to which one group's perception of economic security can be another's idea of insecurity.[15] In view of the high degree of legitimacy attached to multilateral coercion, projects of global economic integration – and the methods of promoting them – are more important than ever before in defining and pursuing economic security.

The sources of economic insecurity discussed so far focus on the international environment, but the impact of globalization on each state's internal socioeconomic stratification should not be overlooked. A globally integrated economy places a higher return on investments in education, as skilled labor and especially technological expertise are increasingly in demand. In states where education is distributed unequally, the greater economic returns available to skilled workers when contrasted with increasing marginalization of unskilled laborers will exacerbate preexisting socioeconomic inequalities. Apart from the obvious economic insecurity that exists for disadvantaged populations themselves, rising inequality could have undesirable political consequences for states such as those in the Gulf that have grown accustomed to social cohesion and peace.[16]

Economic Security and National Development

The tension between globalization and economic security is particularly important in the Gulf context since it raises the question of whether economic integration is consistent with addressing core development issues. In other words, assuming that the risks posed by integration can be effectively managed at a low cost, does this imply that globalization is a substitute for national development strategies? A more compelling economic security framework for the Gulf countries must address not only the challenges faced in the contemporary globalizing world economy as outlined above, but also the development challenges that are specific to these countries.

What development challenges are the Gulf countries facing in the next two decades? Job creation is arguably the most important challenge and it has been the subject of intense policy focus and concern in every Gulf country in recent years in the light of high levels of youth unemployment.[17] According to recent projections by the World Bank, the national labor supply of the GCC economies will grow by an annual average rate of around four percent in the next two decades (See Figure 2.2). Creating jobs for the millions of new entrants into labor markets implies more than a doubling of the current level of the employed national labor force by 2020, aside from the close to half a million unemployed nationals at present and any future inflows of expatriate workers.

Figure 2.2

Labor Supply Growth of Nationals in GCC States (1975-2020)

(annual percentage)

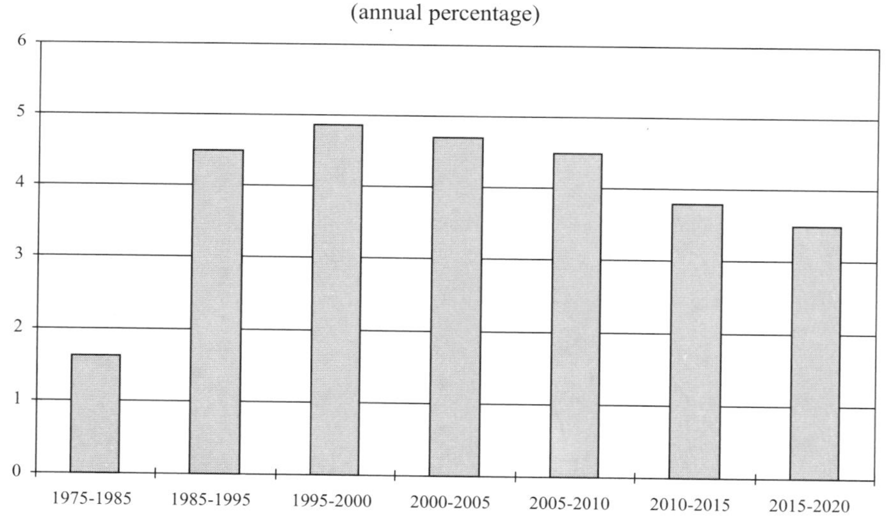

Source: *Unlocking the Employment Potential in the Middle East and North Africa? Toward a New Social Contract* (Washington, DC: World Bank, 2004) and also author's calculations.

However, there is a qualitative dimension to the challenge of job creation in the Gulf countries as well. There is the additional requirement that any new jobs created should entail high wages and benefits in a manner consistent with the rapid expansion of educational attainment, high expectations by young male and female colleagues, and the need to raise worker productivity in the region.[18] Consider, for example, the educational profile of the labor force in the Bahrain and Kuwait where the average years of schooling attained by adults aged 15 and above increased from 1.0 and 1.4, respectively, in 1960 to 6.1 and 7.0 years in 2000. The current levels are closer to standards in Eastern Europe and East Asia than elsewhere in developing countries.

The same constraint applies when considering the effect of expectations by these new entrants into labor markets. The perpetuation of employment guarantees in government hiring and mismatched wage expectations resulting from generous public sector compensation policies have ensured continued queuing for public sector jobs, especially among educated first-time job seekers. In the GCC countries, public–private segmentation resulting from wage and non-wage advantages for nationals in the public sector are further reinforced by distinctions in employment between nationals and expatriates. Private sector wages are considerably lower in countries that rely on foreign laborers not covered by social protection legislation and benefits.

While global economic integration may help the Gulf countries address the challenge of job creation, it is unlikely that this alone would solve the problem. First, the most globally integrated sector in many of these countries – in the sense of competitiveness – is hydrocarbon exports. Yet, even if oil prices remain high, the hydrocarbon industry is highly capital-intensive and the number of workers employed is quite small as a fraction of the workforce. Second, other manufacturing exports, with the possible exception of petrochemicals, face intense competition from China and India. Bahrain and Dubai have created successful service sectors that compete internationally but this required more than simply embracing globalization.

This is not to suggest that labor force pressures could be addressed in these countries by the traditional engine of job creation and public sector employment. If the Gulf countries were to replicate their job creation record for nationals over the past decade during the next ten years, existing projections indicate that unemployment rates could rise from their current levels. The reasons for this are straightforward. While the public sector may continue to be a source of employment for nationals, it is highly undesirable that it will remain a leading source of job creation. Fiscal constraints and low worker productivity imply that any expansion in public sector employment will come at an increasing economic cost and may not be sufficient to absorb the lines of unemployed and new graduates queuing for government employment.

National Development Strategies for the 21st Century

The foregoing discussion suggests that future economic security in the Gulf requires not only global economic integration and proactive strategies to manage its risks, but also the integration of this agenda into broader national development strategies. Otherwise, countries run the risk of amplifying the costs of adjustment and dislocation that may arise, not to mention the neglect of other domestic policy priorities. It turns out that these two policy objectives – globalization and national development – are not only potentially complementary to one another but also mutually reinforcing. As such, together they form the pillars of economic security. In the ensuing discussion, these linkages are outlined using the challenge of job creation as a starting point.

The persistence and, in some cases, the continued rise of unemployment in the past decade along with the projected flows into labor markets in the future has called into question the merits of selective reform programs that are typical of countries in the Middle East.[19] Whether in the domain of relaxing labor market regulations, rationalizing the size of the public sector or nationalizing their labor forces, the record of the last decade suggests that direct interventions in Gulf labor markets alone will be insufficient to address the employment challenge. The same is true for other selective measures involving the privatization of public sector enterprises, the creation of free trade zones or the launch of international joint ventures. Individually, these do not constitute a development strategy for the Gulf countries.

Instead, recent proposals from within the region and abroad have converged on the need for a comprehensive approach to reform in the future. This emerging conventional wisdom suggests that the region's economies should address a set of long-standing policy and institutional challenges to complete three fundamental, interrelated realignments within their economies:[20]

- From public sector to private sector dominated economies, by reducing the barriers to private activity while creating regulatory frameworks that ensure that private and social interests coincide.
- From closed to more open economies, by facilitating integration into global commodity and factor markets while establishing safeguards for financial stability and social protection.
- From oil dominated and volatile to more stable and diversified economies, by making fundamental changes in institutions managing oil resources and their intermediation to economic agents.

In addition, for the region to complete this long-standing transition, rapid progress is needed in educational reform, gender equality and better governance. In other words, the new conventional wisdom calls for an acceleration of the reform agenda that was proposed in the early 1990s but left unfinished.

The Progress of Transition in the Gulf

Compared to the rest of the Middle East, the Gulf countries have made the most progress in terms of improving the business environment and integrating into the world economy. Not surprisingly, Bahrain, Qatar, Oman and the UAE topped the rankings in the 2005 Arab Competitiveness Report.[21] These rankings were made on the basis of the Growth Competitiveness Index (GCI), which measures the ability of economies to achieve sustained economic growth over the medium to long-term (See Figure 2.3). The components of the GCI – technology, institutions and the macroeconomic environment – are easily mapped into the development strategy outlined above. Hence, maintaining competitiveness in the age of globalization is consistent with the goals of national development and hence serves economic security.

Figure 2.3
Growth Competitiveness Index (2005)

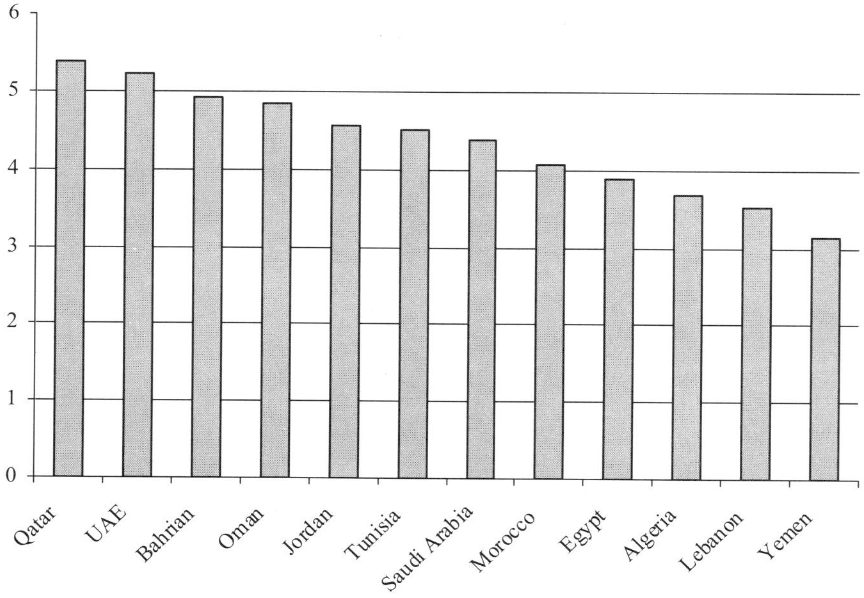

Qatar, UAE, Bahrian, Oman, Jordan, Tunisia, Saudi Arabia, Morocco, Egypt, Algeria, Lebanon, Yemen

Note: Scores range from 1 to 7.
Source: Arab Competitiveness Report, 2005

In particular, the smaller Gulf countries have dynamic service sectors, with substantial recent openings to foreign and domestic private investment in water, power and telecommunications.[22] Independent power projects are being enhanced in Abu Dhabi, Dubai, Oman and elsewhere. Water sectors are also being opened to foreign and domestic investment. Corporatization and privatization measures are being advanced, along with pricing reforms including those that target expatriates. Foreign ownership of real estate has been opened up in Bahrain and offshore-type real estate liberalizations are proceeding in Dubai. Foreign investors are also being pulled into emerging regional hubs for information and telecommunication. No wonder that the Gulf countries are attracting the bulk of foreign direct investment (FDI) coming into the Middle East.

Regulatory improvements are also affecting the financial sector. In view of the approaching deadlines for compliance with WTO conditions and the GCC agreement to form a monetary union by 2010, the Gulf economies have begun to gradually remove barriers to entry and promote competition in their banking systems.[23] These changes have been particularly dramatic in Kuwait and Saudi Arabia, which have traditionally maintained restrictive entry policies. Consequently, the region has witnessed in the last three years a flurry of cross-regional mergers and acquisitions and branching activities in the banking sector. By allowing the creation of pan-GCC institutions, policymakers hope their banking systems will become more efficient and meet the competition from global multinationals.

Two decades ago, most standard infrastructural services in the GCC remained in the domain of state bureaucracies. Since the early 1990s, governments have moved to liberalize the provision of these services through changes in ownership and management structure as well as the introduction of market competition. Other efforts to liberalize infrastructure services were pursued by removing barriers to entry. Line ministries previously in charge of services have been restructured, commercialized or sold off in whole or in part. These reforms required significant changes in the legal framework as well as the introduction of regulatory bodies to supervise pricing and quality issues.

However, the public enterprise sector remains extensive with most large, non-oil industries in public hands and heavily reliant on direct and indirect subsidies, including low-cost loans. As a result, states continue to account for a large share of output and investment in the economy. New firms, particularly small firms, face barriers to entry, both in time and cost of administrative approvals and in securing start-up and operating capital. For example, the minimum starting capital (as a percent of per capita income) in Saudi Arabia and the UAE is around 1000 and 400 percent, respectively. Similarly, the average number of procedures required to enforce a contract in Kuwait and Oman exceed the average in Eastern Europe, Southeast Asia and even the Middle East.

The importance of the public sector is most visible in the Gulf's labor markets. Although the contribution of government employment growth to total employment growth has been low since the private sector creates 90 percent of all jobs, the low share of government employment reflects the decline in expatriate hiring and conceals the continued importance of government employment for nationals. Thus, the public sector accounted for nearly all net job creation for nationals in Kuwait and 40 to 45 percent in Bahrain and Oman. Most of the jobs in the private sector were in relatively low-skill and low-wage sectors and, in most cases, nationals have preferred the slow-growing public sector (See Figure 2.4).

Figure 2.4

Growth in Total Employment of Nationals by Sector (percent)

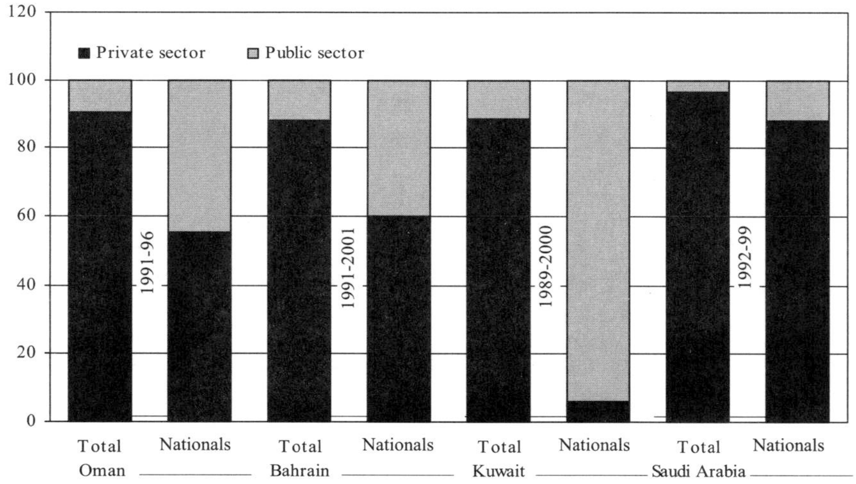

Source: M. Girgis, F. Hadad-Zervose and A. Coulibaly, "A Strategy for Sustainable Employment for GCC Nationals," Working Paper, World Bank (2003).

The GCC countries have embarked on deeper reforms to facilitate greater integration with the global economy. They have already established a US$335 billion customs union and largely eliminated barriers to the free movement of goods, services, national labor and capital within the GCC. To make the customs union work and expand the gains from the planned currency union, further progress will have to be made in establishing common customs rules and procedures, harmonizing technical and regulatory procedures, increasing transparency and minimizing administrative barriers. Such an expanded scope of regional cooperation should facilitate intra-group trade, negotiations with the WTO and increased FDI flows. More importantly, it will enhance the ability of countries to collectively reduce financial risks and manage external shocks.

Where structural reforms to encourage the private sector and promote trade and investment integration have taken place, the end result has been greater economic diversification.[24] Despite the extent of oil wealth and its state control, Bahrain and the UAE at present enjoy the lowest shares of oil exports to total exports and oil revenues to total government revenues (See Figure 2.5). In the UAE, much of the average real GDP growth of seven percent in the last decade was created through rapid diversification in non-oil sectors. At present, the non-oil sectors account for close to 75 percent of GDP and over 50 percent of exports. As a result, the growth of employment in these diversified sectors in the UAE has averaged around eight percent, and the country has the lowest unemployment rate in the region.

Perhaps the most pressing area for reform in order to ensure more stable and diversified economies is that pertaining to oil-resource management. Oil revenues constitute an average of 70 percent of total government revenues and hence, they exert a direct and large effect on fiscal stability. While these revenues finance expansions in physical and human capital and subsidies of goods and services, they also make countries vulnerable to volatility in international oil markets, leading to pro-cyclical fiscal policy and magnifying the effect of oil booms and busts on the non-oil sectors.[25] Moreover, as documented in the literature, large oil rents lead to an appreciation in the real exchange rate and could undermine the external competitiveness of tradeable sectors.

Figure 2.5

Economic Diversification in the GCC (1998–2003)

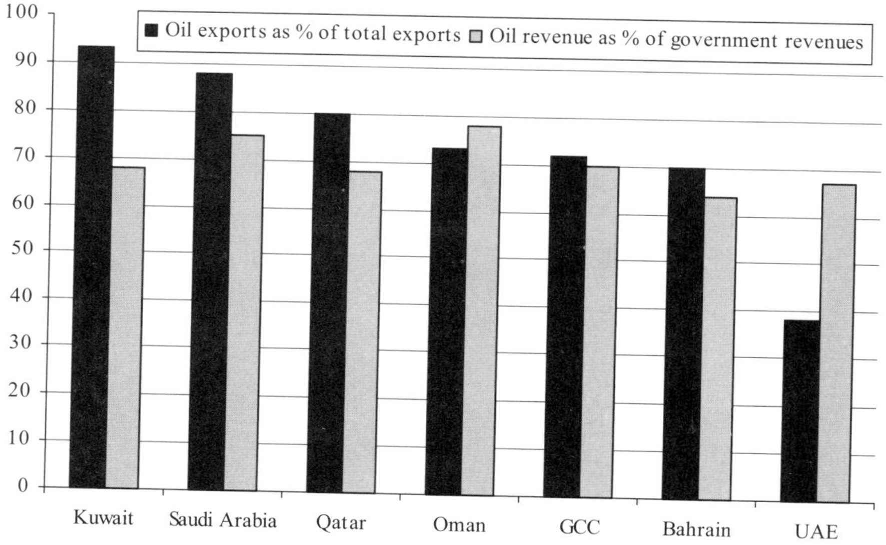

Source: Fasano, Ugo and Zubair Iqbal. "GCC Countries: From Oil Dependence to Diversification," International Monetary Fund (2003).

In the absence of specific fiscal rules or institutional arrangements governing the management of oil revenues, oil price volatility will be reflected in the pace of economic activity, as was the case in the unsustainable expansion and sharp correction to stock markets in the Gulf witnessed in March 2006.[26] Stabilization funds have been developed to lower the impact of volatile resource revenues on the government and the economy by smoothing expenditures flowing to the budget? saving during a revenue windfall and spending during a price slump. The existing formal and informal funds in the Gulf are not as transparent as those in other regions including the rest of the Middle East. Their limited effectiveness may be seen in the acceleration of government spending in the last year following the unexpected increase in oil prices.

Progress in Supporting Re-Alignments

Apart from the goals of expanding access to education and deepening the skill base in the economy, the Gulf economies need to address concerns about the quality of educational outputs and skill mismatches since they affect today's labor market outcomes and future progress with the three transitions explored above. Tackling these mismatches is partly a matter of improving the underlying quality of education and training curricula. There are few performance indicators for schooling in the GCC, and thus little measurement of school quality. Only Oman has attempted to assess the performance of its own education system relative to international learning standards. Educational reform also requires reforming the rigid and centralized management systems that operate in isolation from the economic environment.

Motivated by the desire to improve quality, most GCC governments have liberalized the tertiary education sector and removed restrictions on local and foreign initiatives, leading to a surge in the number of private colleges and universities. Despite the rapidly increasing number of private universities in the Gulf, two issues remain unresolved. The first relates to establishing quality control and accreditation measures for the private universities. Efforts in this regard have proceeded on two fronts: first, by establishing national accreditation councils, and second, through collective efforts to establish regional councils in charge of coordination, accreditation and quality control. So far, Saudi Arabia, Oman and the UAE have established national councils.

At the global level, innovations in information and communication technologies have transformed how goods and services are produced and what products are created. Information and communication technologies are the foundation of the new "knowledge economy" and are associated with the employment of more skilled workers. Empirical studies of the relationships among openness, human capital and total factor productivity (TFP) have found that outward-oriented economies experience higher growth through higher TFP as a result of the positive effect of openness and through a higher impact of human capital on growth. Dubai's stellar growth record may be directly related to its emergence as both an important regional hub for trade and information technologies.

Broad-based economic prosperity in the future requires greater attention to factors driving gender inequality.[27] The Gulf countries have made tremendous progress in closing the gender gap in the education of their female population. In 1980, the average number of schooling years for females aged 15 and above in Bahrain and Kuwait was around two years. Between 1980 and 2000, this indicator of human capital accumulation in the two countries grew at the fastest rate in the Middle East, with the average years of female education in Bahrain and Kuwait in 2000 exceeding six years. As a result, female educational attainment in the small Gulf countries has surpassed the levels of most developing regions.

Notwithstanding this progress, female labor participation among nationals in the Gulf remains substantially below what would be predicted on the basis of education. Thus, substantial economic assets in these societies remain underutilized. In addition, the improvements in socioeconomic indicators have not translated into greater political empowerment for women. Thus, the presence of women in political arenas and their influence on public policy are more limited. Just as in other areas of the Middle East, these low rates of female economic and political participation are increasingly incompatible with the requirements of national development and by extension, the demands of economic security.

Change in this domain can be facilitated by taking the actions necessary to allow for greater participation of women in the labor force and in the public sphere in general. In some countries strict gender norms, which underlie a range of civil, commercial, labor and family laws and practices in the region, discourage women from entering the workforce. Allowing women to become greater economic assets requires reviewing and revising policies and laws, including labor legislation, to eliminate artificial discrepancies that raise the cost of hiring women relative to men. Even integration into the world economy, as evidence from the Middle East suggests, has the potential of feminizing the labor market by opening up sectors that were previously closed to women.

Nevertheless, the challenge of a successful transition to a market-led economy in the age of globalization is primarily a challenge of governance. It is no secret that political systems in the Middle East including the Gulf countries are characterized by limited inclusiveness and accountability especially in comparison to the rest of the world. Thus, enhancing governance extends beyond improvements in bureaucratic performance and reductions in transaction costs that discourage private investment. In particular, efforts to advance reforms hinge on the credibility of government and the capacity of state institutions to manage national development strategies and embrace globalization under conditions of economic volatility and regional insecurity.

At present, governments in the region are handicapped by the limits of institutional structures organized to support redistributive and interventionist policies and the difficulties faced by such institutions in adapting to new tasks, new policy demands and new regulatory environments in the age of globalization.[28] Such instruments are necessary to establish and maintain conditions that promote socially equitable strategies of market-oriented economic growth. Thus, governance reforms are also essential to permitting governments to credibly articulate and realize economic security. The tasks associated with this aim demand a degree of government initiative, creativity and competence that must be cultivated aggressively throughout the region.

To move forward, governments themselves need to link economic performance to political performance and the quality of governance. They need to create rule of law mechanisms to ensure their own accountability and transparency, including in budgeting and fiscal policy, to enable citizens to scrutinize government performance and hold officials accountable for their actions. No less important, governments need to improve the quality and quantity of data on which effective policymaking depends by building more effective infrastructures for data collection and analysis. As noted above in the context of capital markets, transparency has become a requirement and a norm for successful integration into the world economy.

Conclusion

In outlining a framework for conceptualizing Gulf economic security in the 21st century, this study notes that previous concepts of economic security in the GCC, which emphasized the security of oil exports and the pursuit of regional integration, are incomplete since they ignored the opportunities offered by the contemporary globalizing world economy. Instead, it argues that integration into the world economy should be coordinated with the national strategies of the Gulf countries to achieve their particular development objectives as well as to maximize the benefits and to manage the risks arising from globalization.

3

The Oil Boom, the Oil Curse and the GCC

Edward L. Morse

The global oil boom, which began at the start of this decade, provides a historically unique opportunity for the Gulf Cooperation Council states to overcome the so-called "resource curse," which has plagued their economies and societies over the past half century.[1] Unlike the oil boom of the 1970s, the GCC countries now have the human resource base and absorptive capacity to set the stage for sustained economic growth when the current cyclical expansion of the petroleum sector ends and the cycle moves toward its next trough. However, there are enormous political obstacles to be overcome while undertaking the necessary steps to achieve this goal.

The dynamics of globalization, combined with the extraordinary economic benefits of the current oil boom, ironically provide the means for the GCC countries to transform their dependency on oil and gas resources into more balanced economic and social structures. Yet, they also create powerful obstacles to the kind of reforms needed for the governments of the region to overcome their endemic resource curse.

This chapter examines both the opportunities and obstacles, pinpointing the minimum benchmarks that need to be set by the GCC countries if they are to seize the opportunities provided by the current boom to ease the adjustment burdens they will have to confront when oil prices eventually subside, global demand falters, and oil and gas incomes shrink once again.

Resetting Priorities

Not surprisingly, GCC foreign policies are based significantly on the objective of maximizing revenues derived from their hydrocarbon resource base and using the strategic global importance of that resource base to influence the major world powers. After all, the extraordinary reserves of petroleum and natural gas that the GCC countries hold are the primary consideration behind treating their governments as near equals alongside leaders of major powers such as Washington, Tokyo, Brussels, Beijing and New Delhi. As recent history has shown, it also provides them an important security umbrella from Washington as well as open and secure sea lanes for access to markets.[2]

Nevertheless, this understandable obsession of the Arab Gulf states with using oil and gas resources for short-term economic and political gain has entailed profound long-term costs. It has stultified or sub-optimized the development, not only of their resource bases but also of the potential diversification of their economies. This situation is readily understood by comparing the development of two companies over the past quarter century.[3] One is a state-owned company from the region, the Kuwait Petroleum Corporation (KPC). The other is a shareholder-owned company, British Petroleum (BP), which was state-owned at the time the first comparison was made.

In 1981, at the height of the last oil boom, KPC and BP were approximately the same size with both having gross sales of around $30 billion. Both companies were challenged by the rapidly changing market conditions of the time, including declining prices and demand. KPC's response to the new market situation was partially dictated by joint decisions within OPEC and inter-governmental agreements to curb output in order to put a floor under prices. Even so, confronting severe competition in the marketplace, KPC was a pioneer in the 1980s among the national oil companies of OPEC in purchasing downstream assets abroad, especially in Europe, in order to secure market share and minimize marketing risk. BP on the other hand, found itself critically "short" of crude oil and extremely "long" on refining capacity, having lost considerable upstream assets in Kuwait, Iran and Nigeria. It was far more challenged than KPC, which retained its solid reserve base. Ultimately, however, BP's response to the challenge proved far more extraordinary.

Eighteen years after the record earnings of 1981, KPC's revenues in the year 1999 were essentially half of their 1981 level. On the other hand, BP's gross sales were roughly ten times greater than that of KPC in 1999. To be sure, there are major differences between the two figures, including the fact that BP, traditionally short on crude oil and long on refining, needs to purchase crude oil for processing and reselling as products. Even so, the spectacular growth in BP's overall volumes stands in stark contrast to the stagnation that beset KPC. The former achieved this sales level by wholesale seizing of globalization opportunities, by investing abroad, by acquiring properties and companies, thereby giving shareholders the assurance that they could maximize their earnings. Its ability to secure these gains was in no small measure assisted by the British government's decision to privatize the company. However, that ability also stemmed from BP's far stronger human resource base in managerial, commercial and technical competencies, which more than made up for KPC's more robust resource base.

KPC shareholders learned two clear "lessons" from their market experience since oil prices reached their low point in 1999. These so-called "lessons" are clearly leading to decisions that go against the logic of globalization and impede the full adjustment of Kuwait to a new global situation. That, in turn, poses an obstacle to Kuwait moving forward during the current oil boom to maximize its own social and economic flexibility with a view to confronting the next cyclical downturn in the oil industry.[4]

The first lesson learned was that KPC could double its revenue by working with other OPEC and non-OPEC producers and cutting production by 10 percent, as the producers did when oil prices fell to $10 a barrel in the winter of 1998-99. As seen in Table 3.1, the result was stunning. In 1980, at the height of the earlier oil boom, Kuwait earned $38.4 billion from oil exports (in constant 2005 dollars). The price crash in 1998 meant the earnings fell to a mere $9.1 billion in real terms and surged to an estimated $39 billion in 2005. This conclusion has led one leading economist analyzing the oil industry to assert that while OPEC countries may have essentially unlimited hydrocarbon reserves, they also have limited incentives to expand capacity. From the perspective of short-run revenues, they would always be better off by cutting production rather than by investing in new capacity, should oil prices fall.[5]

Table 3.1

OPEC Net Oil Export Revenues at a Glance (1972-2007)

Actual and Estimated (in Nominal and Constant US Dollars)

	Nominal Dollars (Billions)				Constant $2005 (Billions)				
	Change								
	2004/05	*2005E*	*2006F*	*2007F*	*1972*	*1980*	*1998*	*2006F*	*2007F*
Algeria	52%	$36.0	$41.6	$41.1	$5.0	$26.4	$6.4	$40.9	$39.7
Indonesia	N.A.	($1.0)	($0.6)	($0.9)	$3.3	$30.4	$3.5	($0.6)	($0.8)
Iran	45%	46.6	50.1	46.5	15.3	26.8	11.2	49.2	44.9
Iraq	31%	23.4	24.9	23.7	5.4	55.3	7.7	24.5	22.9
Kuwait	41%	39.0	44.1	41.1	10.3	38.4	9.1	43.3	39.7
Libya	52%	28.3	31.2	29.9	10.9	45.5	6.7	30.7	28.9
Nigeria	40%	45.1	52.7	51.1	7.8	48.8	10.0	51.7	49.3
Qatar	28%	19.1	23.3	23.0	1.7	11.0	3.9	22.9	22.2
KSA	49%	153.3	162.0	150.2	17.2	213.6	36.9	159.1	144.9
UAE	44%	45.6	53.0	52.2	3.9	38.5	11.5	52.1	50.4
Venezuela	32%	37.7	39.4	37.2	11.3	37.2	13.6	38.7	35.9
TOTAL	**43%**	**473.1**	**521.9**	**495.2**	**92.0**	**571.8**	**120.7**	**512.5**	**477.8**

Note: NA = not available; E = estimates; F= forecasts.

Source: *OPEC Revenues Fact Sheet* (Washington, DC: Energy Information Agency, US Department of Energy, January 2006).

The second lesson learned – which is embodied in the philosophy of KPC's shareholder, the government of Kuwait and in the similar shareholder philosophies of other OPEC national oil companies – is to be wary of capacity expansion. The main concern over capacity expansion is shown in the obsession of most OPEC governments with regard to market security. The question they raise is: If they undertake investment in capacity expansion, who will provide the assurance that demand will continue to grow? This theme has dominated discussions by producer countries in recent years. It highlighted the interventions of OPEC producer governments, for example, at the meeting of the International Energy Forum in Saudi Arabia in the autumn of 2005. At that meeting Ali Al-Naimi, the Saudi Minister of Petroleum and Mineral Resources made the following statement:

> We would ask the consumer nations to give us a demand road map, because as producers we do not want to build facilities that do not have demand for their production.[6]

In a world in which demand is impacted by seasonal factors, economic conditions, price levels and a host of other factors, it is hard to understand how or why a responsible party could expect a clear "road map" of future demand. However, it does offer an explanation for the inclination by producer nations to go slow in building new production capacities.

These two so-called "lessons of history," which dominate the thinking of the government owners of national oil companies, have had two significant adverse impacts on the GCC countries. The first impact is associated with the behavior of the state companies themselves. The second impact pertains to the international position of the governments that own the state companies and in turn, on the international arena that characterizes the trade and investment rules associated with the petroleum sector.

With respect to the national oil companies (NOCs) themselves, past lessons have made them far less adaptive to rapidly changing conditions facing the industry than the international majors and independent companies. Certainly, they retain control over huge resource bases. However, as the above-mentioned example about KPC and BP reveals, they have also brought *lower* returns to their shareholders than have the IOCs. In turn, this has impeded economic and social development within their societies. The history of per capita export revenues in OPEC countries from 1981 to 2000 makes this point especially clear (See Figure 3.2). In constant 2005 dollars, OPEC per capita revenues peaked at about $1,800 in 1981 and fell to a low point of $250 in 1998. Even with the oil price rise after 1999, per capita oil export revenues remained under $1000 in 2005? four times higher than in 1998, but only slightly more than half of their 1981 level. The GCC countries may have fared better than other OPEC members, with per capita revenues in some cases now standing at a historical high. However, that high point is barely above the level that prevailed in 1980-81.

Figure 3.1

OPEC Per Capita Crude Oil Export Revenues (1971–2007)

Actual, Estimated and Projected (in constant US dollars)

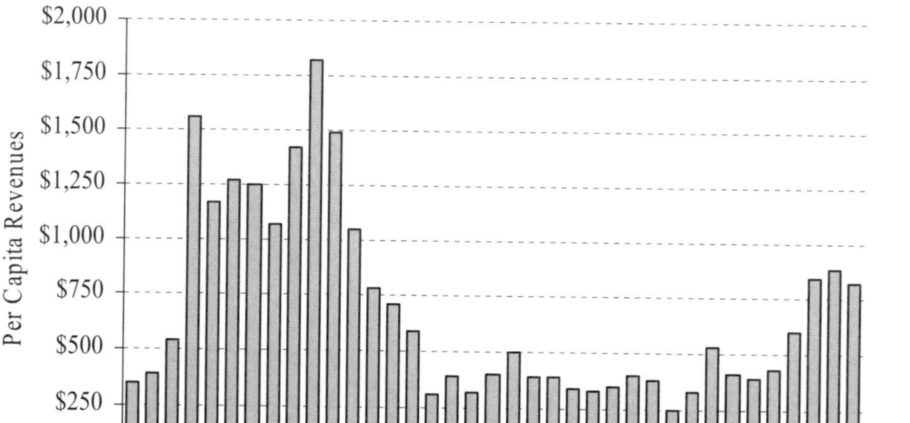

Source: Energy Information Administration, US Department of Energy.

When it comes to the international implications, the results are equally stark, and it is worth focusing some attention on this issue.

Winners and Losers[7]

Oil has become the most glaring exception to the general rules of trade and investment underpinning globalization. The widely recognized benefits of globalization are associated with the rules that encourage the international flow of goods and capital. In trade, these rules involve the breaking down of barriers, whether in the form of tariffs or quotas or an array of non-tariff barriers. In both trade and investment, they involve enshrined principles of reciprocity and non-discrimination. These rules aim to depoliticize and stimulate flows of goods, services and capital to enhance global growth and welfare, thus encouraging both individuals and companies to enjoy the benefits derived from maximizing gains for all parties concerned.

These principles are considered win-win propositions, with individuals, companies and societies benefiting from ever-enlarging globalization. Of course, there are frequent bumps along the way, due to dislocations affecting competitiveness, employment and peripheral national goals. However, the benefits are generally recognized as far outweighing the costs of enhanced interdependence.

Not so with oil. OPEC, founded some 45 years ago, represents a glaring exception to the world of globalization. According to some analysts, it embodies and enshrines certain principles that may be considered discriminatory.[8] They argue that rather than removing economic instruments as tools of foreign policy and thereby defusing tensions, it positions oil as a centerpiece of foreign policy for its member states. Consequently, oil can be sold on a discriminatory basis and the oil trade becomes a matter of state-to-state relations and becomes linked to other foreign policy goals. Typically, OPEC producers sell crude oil and products on a destination basis, refusing to allow buyers to resell outside the region to which the oil is sold, thus hindering the development of spot markets and greater market transparency. Barriers to investment are preserved and protected as governments freely discriminate against foreign investment or totally ban it, providing their companies with unfettered, uncompetitive monopolies in the exploration, production and sale of oil.

Consequently, unlike the rest of the global economy, the rules in the oil sector are set up for win-lose or zero-sum outcomes: "My advantage is your loss. How can I gain and push the burdens of adjustment onto you?" Ultimately, however, the winners cannot be too harsh on the losers. If prices rise too high, losers will take countervailing actions that punish the short term winners, sooner or later. However, when prices are too low, the producers have discovered that significant leverage can be found to push the burden of adjustment on the buyers. The zero-sum nature of the international petroleum sector is not always at the forefront. Yet this nature emerges with a vengeance when oil prices fall either far below or rise well above historical norms.

It was not always this way in the oil sector, even if it may be argued that for much of the past century the "hidden hand" of the market was guided by the sometimes not-so-hidden regulator, whether it was the Texas Railroad Commission (TRC), which pro-rated the bulk of US output to shore up prices from collapsing, or the "Seven Sister" oil companies, which kept prices steady or from falling, expanding capacity at a balanced pace and holding it back supply when markets softened. Yet neither the Texas Railroad Commission nor the Seven Sisters discriminated among buyers or destinations.

As developments unfolded from the early 1960s to the mid-1970s a decisive break occurred. First, OPEC was created in order to bolster producer governments in their efforts to secure a higher share of revenues and to use the companies that had invested in and developed their oil production to pressure Washington to reduce its import controls. These controls had caused the oil producers to lose access to the market in the United States (which incidentally, was blatantly protecting its own industry) and resulted in a price reduction in the rest of the world.[9] Second, OPEC itself came to symbolize an emerging "North vs. South" struggle within all commodity markets as commodity producers tried to forge a "New World Economic Order" by using their collective clout to force change on the industrialized world.[10] Third, and by no means least important, Washington encouraged certain OPEC producers – notably, but not only Iran – to seize the initiative in raising prices and enhancing the role of the government in the upstream sector as a means of strengthening these countries in efforts designed to curtail expanded Soviet influence in the Middle East as part of the Cold War struggle.[11] At a later period, when Washington and Riyadh had comparable concerns about the Soviet Union's ability to expand abroad as a result of higher prices, there is evidence that the two countries cooperated at the highest levels to reduce prices well below prevailing levels, including the crash to $10 in the mid-1980s. For Washington, the motivation might have been to deprive Moscow of the means to fight insurgency in Afghanistan while also maintaining various subsidies for the Warsaw Pact. For Riyadh, in addition to Afghanistan, there was an incentive to deprive Iran of financial resources at a time when it was warring with Iraq and spreading its brand of Islamic fundamentalism in the Arabian Peninsula.[12]

Thus the prevailing domestic factors in key Gulf oil producing states were reinforced by the international framework, which came to characterize the global petroleum sector.

Today's Petroleum Boom: New Opportunities

A striking economic expansion throughout the GCC countries has accompanied the rapid increase in base oil prices since 2002. Although this economic boom may have been sparked by the enormous increase in government spending that accompanied rising government oil revenues, the truly extraordinary character of the economic expansion has been the thus far sustained effect in non-oil sectors. This has been made possible largely by the changes in the economy and human resource base of the GCC societies, which, lacking such a base during the last oil price-led economic expansion, could not sustain economic growth when oil prices and government spending fell in the 1980s.

Now, with the prospect of sustained high oil prices accompanied by sustained high government spending, there is a good chance that secondary effects can foster truly diversified economies within the region. These secondary effects include investments in real estate, tourism and financial services in particular, as well as the opening of previously closed sectors to private investments through the public offering of shares in both new corporations and privatized companies. The accelerated economic expansion in certain parts of the Gulf – perhaps best exemplified by Dubai, where the economy has currently ceased to be oil-based – raises the question of whether the GCC countries can now set a "beyond oil" basis for regional economic growth.

The GCC data on aggregate indicators show just how remarkable the recent economic expansion has been (See Table 3.2). GDP per capita has grown by more than 40 percent and the non-hydrocarbon part of the GDP has grown by a sustained 9.4 percent. The liquidity growth in the private and public sectors has seen a rapid expansion in intra-regional capital flows, with investments concentrated in real estate, local equities and tourism. The local capital markets and banking system have grown substantially as well, assisted by government-induced reforms and the establishment of regulatory frameworks modeled on those of the OECD.

Part of the GCC economic diversification has been stimulated by sustained US government spending in the region, supported by the growth in free trade agreements with the United States. This spending has come both from the larger US military presence in the area and from projects launched in the region via US government spending for Iraqi reconstruction. Direct expenditures in Iraq have been running at about $100 billion per year, of which more than $20 billion involves procurement, with a significant amount of more than $50 billion for operations, logistics and maintenance spent in the GCC countries (not to mention the recreational money spent by US military forces in the area).[13] Over and above these Iraqi expenses are the costs associated with the deployment of armed forces in the Gulf region, which even before the

Gulf War was estimated at some $60 billion annually. This was before the creation of new bases in Qatar and Oman and the expansion of facilities in Kuwait and Bahrain.[14]

Table 3.2

GCC Aggregate Economic Indicators (2003–2005)

INDICATOR	2003	2004	2005e
Nominal GDP ($billion)	**399.2**	**469.3**	**585.7**
GDP per capita ($ '000)	**12.1**	**13.7**	**16.5**
Nominal GDP (% change)	**15.1**	**17.6**	**24.8**
Hydrocarbon GDP	24.5	33.0	46.2
Non-hydrocarbon GDP	10.3	8.5	9.5
Current Account Balance ($ billion)	**51.7**	**92.8**	**180.7**
% GDP	12.9	19.8	30.8
Oil exports ($ billion)	146.5	195.7	291.4
Gas exports ($ billion)	10.1	13.4	19.6
External debt ($ billion)	77.9	87.7	99.2
(% GDP)	19.5	18.7	16.9
Fiscal Balance ($ billion)	**31.4**	**64.4**	**123.2**
% GDP	7.9	13.7	21.0
Fiscal revenue ($ billion)	162.4	212.4	296.6
Fiscal expenditure ($ billion)	131.0	148.0	173.4
Memoranda			
Oil price ($ Brent)	28.5	38.0	54.1
Crude oil production (mbpd)	14.9	15.5	16.2
Natural gas production (mboepd)	3.0	3.2	3.6

Note: e = estimate

Source: Institute for International Finance (IIF), "Gulf Cooperation Council Countries," Regional Report, July 29, 2005.

Some regulatory changes have also been US-inspired, such as the anti-money laundering efforts sparked by the post-9/11 desire to curb terrorism financing. These efforts have been buttressed by various campaigns led by the International Monetary Fund (IMF) and the Institute for International Finance (IFF). Corporate and banking transparency and good governance efforts are being pursued by governments in the region as well as by international governmental and non-governmental organizations, in efforts to ensure access to the international banking system and promote property rights essential to the building of financial institutions in the region.[15] Additional economic benefits have also come from the regional impact of WTO membership for virtually all the GCC members, including Saudi Arabia, and the expansion of free trade agreements between individual GCC states and the United States, including Bahrain, Oman and the UAE. While it is too early to find clear measures of the impact of these agreements on increasing trade between these countries and the United States, there is ample evidence that partly as a result of these agreements, the countries involved have accelerated their domestic reform agenda and implemented legal processes to bolster the role of law to protect property rights including ownership of shares and intellectual property.[16]

In short, the GCC economic boom, although fueled by rising oil prices, has also been fostered by the economic benefits of US government spending and the extraordinary economic liberalization that has been gripping the region.

After nearly two decades of persistent stagnation, most GCC members (with the notable exception of Saudi Arabia), have plunged wholeheartedly into the globalized international economy and have begun reaping the associated benefits. A critical question for government officials in the region is what measures can be undertaken to ensure that structural change in regional economies remains adequately deep-rooted to enable continued growth, once the current phase of rapid expansion in the petroleum sector ends and the oil price boom tapers off, reaches a plateau, and eventually reverses itself. Two critical sets of issues must be confronted. First, what are the lessons to be learned from the last contraction period in the oil sector? And second, what can be done to ensure sustained economic expansion in the future?

Lesson from History

There are two very clear lessons from the past, both of which are difficult to recall at the present time, when economic exuberance prevails. Yet, both are critical to remember if the GCC economies are to prepare themselves for sustained growth into the 21st century. In all likelihood, some action is required to ensure growth and political stability once the oil boom ends. The first lesson is that all economic expansions eventually come to an end. The second is that low prices are unfriendly to economies that are overly dependent upon single commodities or crops.

It should be evident to the GCC governments and citizens (as well as their large expatriate communities) that all oil booms must inevitably come to an end. When they do so, prices reach a critical turning point, which is sharp, unexpected and does not reverse itself for a long period.[17] The last petroleum boom, which occurred during the 1970s, was a period in which nominal oil prices increased by close to 2000 percent and reached a peak in 1980. This was followed by a long period of exceptionally low prices by the standards of the late 1980s. When the turning point arrived, there was disbelief throughout the producing world that lower revenues and lower incomes were becoming permanent features of the economic landscape. Individual policies and approaches adopted by the GCC countries and their collective behavior within OPEC (for those GCC members belonging to OPEC), amounted to a short-sighted strategy of adopting temporary measures to protect themselves during this period of low oil prices, in the vain hope that these low prices would soon come to an end. It took twenty years of that misguided policy to eventually reap its reward: twenty years of economic pain, protracted delay in adopting adjustment policies, and lost opportunities.

The second set of lessons is associated with the effects of a sustained period of low prices. What happens within oil producing countries? What happens in the relationships between oil producing countries and other countries that are not so heavily dependent upon oil revenues?

By now these lessons should be clear.[18] First, with respect to the lessons for producers, low prices imposed significant burdens:

> ...economic hardship in oil-producing countries, including declines in oil revenue, budget deficits, budget cuts and cancelled projects, borrowing and debts, deterioration in the balance of payments, negative economic growth, currency devaluations and political unrest.[19]

The burdens on individuals is reflected in the sharp decline that occurred in the 1980s and 1990s in per capita income, especially as lower or stagnant revenues confronted significant population growth, and lower government revenues and expenditures confronted higher unemployment. Second, with respect to international relationships, low prices and a downturn in oil and gas industry activity increased the gaps between producing countries and the industrial world, increased friction among oil producers, decreased the importance of oil producing countries to key industrialized countries, thereby lessening their international importance, and enabled oil-importing countries to impose high taxes on oil products, further depriving the oil producing countries of revenues that were deemed to be theirs.

These "lessons" of the recent past are largely responsible for the adoption of defensive policies by most Middle East producers. It led them to be cautious about expansion plans, especially in the upstream sector, and to focus on problems of "security of demand."[20]

Planning for Sustained Growth

One characteristic of the GCC economies that has been associated with the current oil boom has been the rapid increase in wealth throughout the region. This wealth increase is reflected in the strong growth in local capital markets, the extraordinary depth of participation within each GCC country in their local stock exchanges, and the rapid pace of real estate development. Clearly, one factor contributing to the rise in GCC local capital markets has been each country's development of a property rights system along with the rule of law for protecting and enhancing those rights. Another factor, on the opposite side of the ledger, has been the inflation growth in both of these property values. The GCC real estate boom appears increasingly to be a bubble and the inflation in local equities is reflected in the sometimes overcharged levels, with price versus earnings ratios rising to levels of well over 50 percent, often to the 100 percent level.

Undoubtedly, one reason why equity and real estate values have become so inflated is the perceived lack of investment opportunities elsewhere.[21] Certainly, in the post-9/11 context, much of the region's new-found wealth is held by people who are concerned about the possible freezing of assets in the United States and other countries.[22] Another factor has been the inability of GCC citizens to participate directly in the current oil boom, due to the monopolization or near-monopolization of oil ownership by the state.[23] Undoubtedly, there are some experiments in change. Within the UAE, there have been initial public offerings (IPOs) of oil and gas firms, even exploration and production companies, which have been incredibly successful, bearing witness to the growing appetite for direct investment opportunities. Overall IPO activity in 2005 in the Middle East region increased by 100% and reached $6 billion.[24] When Dana Gas, the first private sector regional natural gas firm in the Middle East had its initial public offering in December 2005, it was oversubscribed by 140 times. Kuwait is now experimenting with sales of downstream assets in product retailing, which is a move in a similar direction.

This seems to be the moment for governments to rethink the strategy of having huge state companies operating in the oil sector. This is not an easy task in a region where the form of government is monarchical rather than participatory, and where the state revenues and the dispensation of essentially free state services have traditionally boosted governmental authority and legitimacy. Undoubtedly there is pent-up demand within the GCC for direct participation in the oil sector, which has made these societies what they are. Of course, partial privatizations could go a long way to slaking this desire. Undoubtedly, with the combination of enhanced property rights and the protective shield of the rule of law in defending those rights, local capital markets can be used to spread that wealth.

There will inevitably be a chorus of objections to the very idea of dismantling these key state assets or doing so too quickly. However, there really should be internal debates within the GCC about the appropriate path. The successful transformation of Dubai – once a producer of over 500,000 barrels per day (bpd) of oil and now producing less than 100,000 bpd – from an oil emirate to a diversified services economy demonstrates a way toward economic diversification.[25] The successful privatizations of once large state-owned enterprises in oil, gas and other energy sectors in the OECD demonstrates that significant benefits can be earned both by state and society by changing the property rights system governing oil and gas. In this connection, the Norwegian example is instructive. Although Norway is a country with a small population, comparable to those of the GCC states, its total hydrocarbon production and exports (natural gas and oil combined) exceeds that of every GCC state with the exception of Saudi Arabia. In the process of privatization, the Norwegian government has discovered that it has lost little control over the oil sector – indeed in many respects it has acquired some of the authority it lacked when it had a huge national oil company – and gained much by separating the regulatory functions over oil and gas operations from the actual ownership. It has benefited from more competition and greater participation by citizens in oil and gas property rights. However, it can still act to curb production if prices fall suddenly and damage the domestic economy.

The Norwegian experience provides a valuable lesson for all the GCC countries. The lesson is that in the current climate of oil sector expansion, there are opportunities to seize win-win solutions. This applies especially to the smaller GCC states, which can integrate their economies into global market mechanisms with much greater ease than the larger GCC members.

Section 2

INDUSTRY TRENDS AND PROSPECTS

4

Refining Challenges: From Crude Oil Quality Availability to Clean Product Demand

Pedro Antonio Merino Garcia

During the last years, some important changes have occurred in the oil sector, both on the supply and demand side that have led to a great tightness in the markets. While oil demand is being driven by significant economic growth (especially in emerging countries), by the increasing weight of the transport sector in final petroleum consumption, and by the new regulatory framework for environmental protection, the supply side is experiencing a slight bias towards heavier and sourer crude oils that, in the long run, will cause a mismatch between crude oil availability and refining capabilities. As always, there is one way to balance crude supply and final product demand requirements and that is to invest in more capacity in the refining industry.

Developments during the period 2002–2006 in terms of deterioration of the crude oil slate produced were not anticipated by forecasters and now the refining sector has to adapt existing production schemes to new conversion units that are able to keep up with the new supply conditions and meet the present demand trends.

In this chapter, the main purpose is to analyze the current challenges facing the refining sector, both from the supply and the demand sides, with a special focus on crude oil quality evolution and final product demand characteristics. The interaction of both demand and supply will have crucial implications for the refining sector, in terms of investment requirements and refining margin levels or profitability.

In the first section, the concept of crude oil quality availability is explained, its current situation and historical trend are described and the projections made by some private consultants and public agencies are analyzed. Apart from these explicit projections, the future crude oil quality according to the expected production levels is also studied.

In the second section, the case for a structural change in demand is evaluated. The acceleration in global economic growth, the increasing share of the transport sector in global oil demand, the expanding role of emerging markets, the resilience of OECD demand and global regulatory change towards more stringent specifications are the elements behind the structural change.

In the third section, refining challenges are discussed. Current bottlenecks are described and the expected investment in refining is analyzed. Due to the uncertainties surrounding new investments, a section is also included on the variables that either encourage or discourage future investment.

Supply: Crude Oil Quality Availability

It is a fact that crude oil markets behave in a global way and that crude oil price is determined by supply and demand interaction. Nevertheless, crude oils produced worldwide show significant differences in physical and chemical characteristics that affect their processing, product yields and prices.

When considering the oil industry in general and the refining sector in particular, future crude oil quality availability is a key issue for discussion, as this factor will be responsible for some changes that the sector will experience.

Many characteristics influence crude oil quality, but aside from the distillation curve, the two most basic and relevant aspects are density and sulfur content. These two properties establish the quality of a crude oil.

In this chapter, the term crude oil quality availability will be understood as the average quality of the crude oil produced worldwide, with regard to density (API gravity[1]) and sulfur content.

Crude Oil: API Gravity and Sulfur Content

In the oil industry, API gravity is the standard measurement for classifying crude oils as being lighter or heavier in composition. Generally, crude oils with API gravity above 34 °API are considered light, those between 27 and 33.9 °API are considered medium and those below 26.9 °API are considered heavy. A remarkable fact regarding density is that in some cases, crude and condensates or natural gas liquids (NGLs) are aggregated so that the average density of the resulting petroleum is lighter than when considering only crude production.

Light crude oils such as Bonny Light, have low heavy residue volume (only 28 percent for Bonny Light) and greater yields on medium and light distillates, thus being more propitious for obtaining high added-value products, such as gasoline and heating oil, from simple refining schemes. On the other hand, heavy crude oils such as Maya generate a high heavy residue volume (around 60 percent for Maya) and have great yields on very low value heavy products like fuel oil, requiring therefore more complex refining schemes including conversion units to obtain the high value distillates. In the middle, medium crude oils such as Urals, still deliver important heavy residual volumes (around 45 percent for Urals).

Sulfur content can vary from 0.1 percent to more than 6.0 percent in weight. The normal values are between 0.2 percent and 3.0 percent. Usually, crude oils are denoted as sweet when their sulfur content is below 0.6 percent and as sour when it equals or is above this percentage.

Crude oil sulfur content affects quality in a negative way. Sulfur must be properly eliminated through certain industrial processes; otherwise it is transferred almost totally to refined products which are submitted to strong environmental constraints. These industrial processes make refining more expensive. Therefore, a sweet crude oil like Bonny Light with 0.14 percent sulfur content is more appreciated than a sour variety such as Maya with 3.33 percent sulfur content (see Table 4.1).

It is important to remark that although most heavy crude oils have high sulfur content, this is not always true and the API gravity of any crude oil is independent from its sulfur content.

Table 4.1

Percent Volume of Residual Distillation for Several Crude Oils

	° API	% Sulphur Content	% Volume Residue
WTI	38.7	0.45	33.20
Brent	38.1	0.39	33.70
Bonny Light	33.6	0.14	28.03
Arabian Light	32.7	1.80	46.70
Urals	31.8	1.35	42.18
Dubai	30.4	2.13	42.80
Arabian Heavy	28.7	2.79	52.60
Maya	21.8	3.33	60.30

Source: Energy Intelligence Research.[2] Compiled by Repsol YPF Research Department.

The deterioration in crude oil quality (with crude oils getting heavier and sourer) has important implications for the refining industry.

To illustrate this point, it is sufficient to consider a decrease of just 1 point on the crude oil API gravity scale and quantify its effects on crude oil residue volume. As seen in Table 4.1, moving from Bonny Light (33.6 °API) to Arabian Light (32.7 °API) implies a shift from 28 percent of residue to almost 47 percent, nearly doubling the residue volume.

Therefore, it is very important to be aware of future crude oil quality availability. Current crude oil production by quality should be an initial short term approximation of the future situation. In the next section, the current and the historical trends in crude oil quality are analyzed.

Current Situation and Historical Trend in Crude Oil Quality

In the following regions – Middle East, Former Soviet Union, South & Central America, North America, Sub-Saharan Africa, Asia & Oceania and Western Europe – the current regional crude oil quality situation could be described as follows:[3]

- In the Middle East, about 80 percent of the production of crude oil is medium and sour.
- In the Former Soviet Union, about 87 percent is medium and increasingly sour.
- In Latin America, more than 90 percent is heavy and since 1996, heavy crude oil is the only kind whose production is increasing.
- Africa is the only region where light crude oil production has recently grown representing approximately 60 percent of total production in the region.

Thus, about 75 percent of the global supply comes from regions producing heavy and medium crude oils.

Figure 4.1

Current Regional Crude Density Slate in Key Production Regions

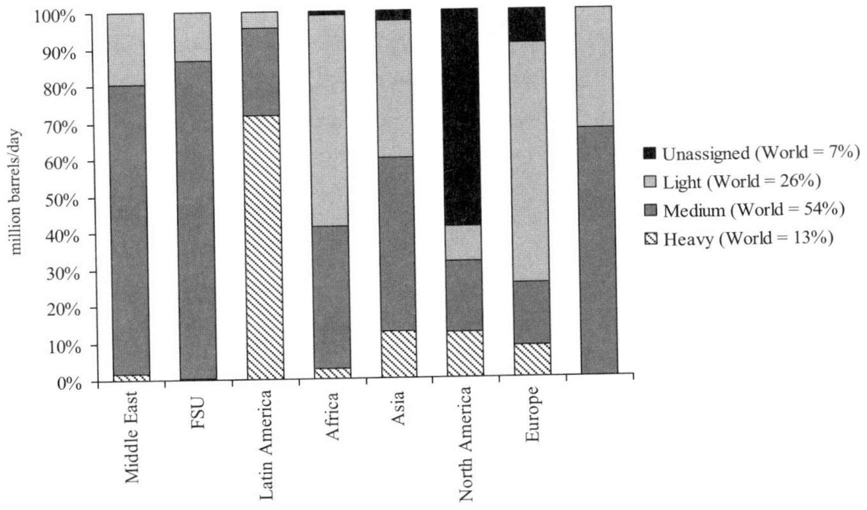

Source: ENI. Compiled by Repsol YPF Research Department.

According to the *World Oil and Gas Review 2006*[4] issued by ENI, more than 65 percent of the 2005 world-wide production was medium and heavy, and more than 60 percent was sour (see Figures 4.2 and 4.3).

In global terms, from 1994 to 2005 the participation of light crude in the production has been reduced from 28 percent to 26 percent, while medium and heavy crude oils participation has been increased from 62 percent to 67 percent.

In volumetric terms and in the same period, heavy crude oils production has grown at faster rates (4.2 percent annual) than medium crude oils (1.8 percent annual) or light ones (1 percent annual).

On the other hand, the share of sweet crude in total production decreased from 32 percent to 29 percent, while the share of sour crude oils grew from 58 percent to 61 percent. In terms of annual growth, the rates are 1.6 percent for sweet and 2.0 percent for heavy varieties.

Figure 4.2
Global Crude Density Slate (2005)

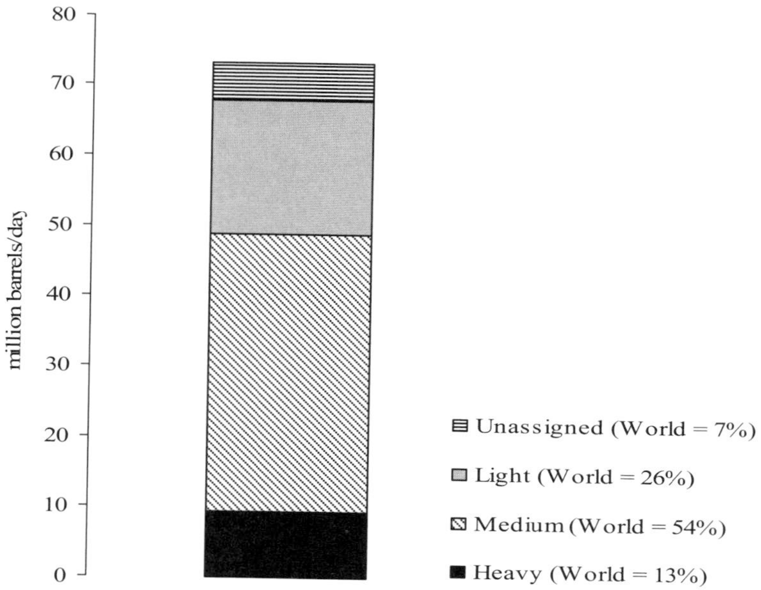

million barrels/day
80
70
60
50
40
30
20
10
0

Unassigned (World = 7%)
Light (World = 26%)
Medium (World = 54%)
Heavy (World = 13%)

Source: ENI. Compiled by Repsol YPF Research Department.

Figure 4.3
Global Crude Sulfur Content Slate (2005)

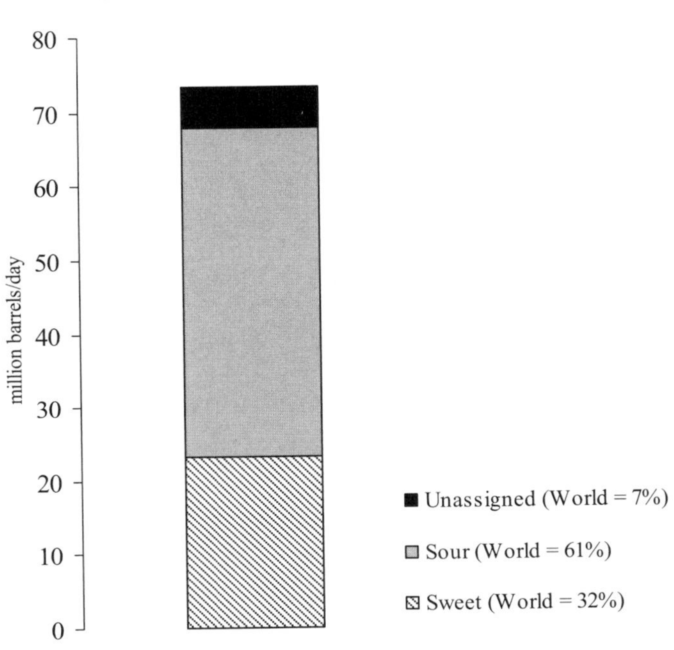

Source: ENI. Compiled by Repsol YPF Research Department.

Regarding the recent historical trend of crude oil quality availability, although different data are handled by experts depending on their respective assumptions, there is enough evidence to conclude that the average of the worldwide crude oil supply quality has been evolving from the lighter and sweeter range, towards the heavier and sourer range.

The trend of the past decade seems to be a continuation of what was witnessed during the 1980s. The evolution of crude oil quality from 1980 to 2000 shows deterioration in terms of both density and sulfur content throughout these years. According to Simmons & Company International[5] (see Figure 4.4), in that period the average crude oil API gravity decreased 2.75 points averaging approximately 32.25 °API and the sulfur content increased 0.5 percentage points, averaging around 1.1 percent.

Figure 4.4
Past Trend in Crude Oil Quality Availability

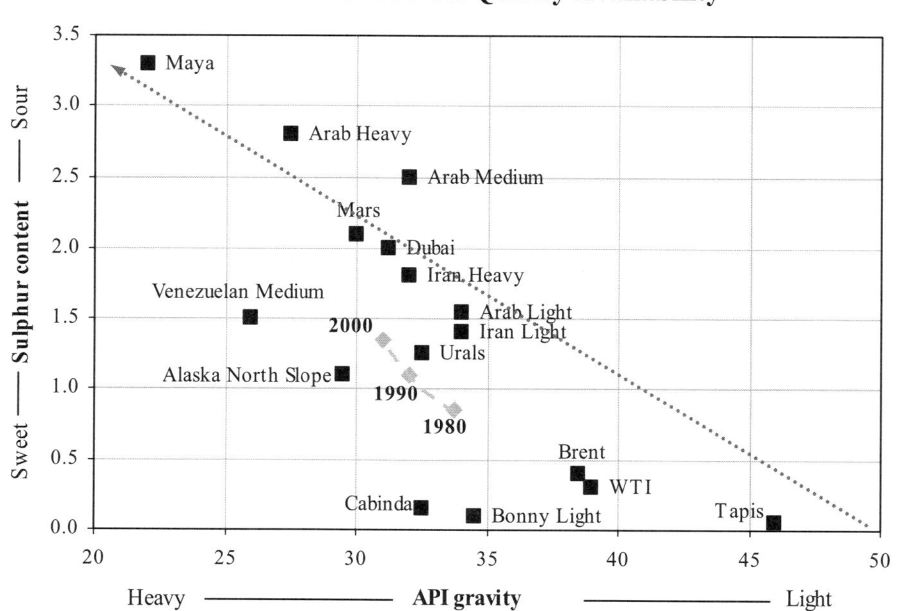

Source: International Energy Agency (IEA). Compiled by Repsol YPF Research Department.

Indicating the same direction, although relying on different figures for API gravity and sulfur content, Wood Mackenzie[6] states that in the last five years, average API gravity has diminished practically 0.5 points, averaging near 33.55 °API. Correspondingly, the sulfur content has increased 0.03 percentage points, averaging 1.2 percent approximately (see Figure 4.5).

It is difficult to determine what are the exact values of API gravity or sulfur content of crude oil production, but the fact is that most experts have identified a trend towards heavier and sourer crude oil production.

More relevant than the actual API gravity is the fact that 67 percent of the crude oil produced in 2005 was medium or heavy compared to only 62 percent in 1994. In this regard, it is especially important to know and to assess the projection for future oil quality.

Figure 4.5
Recent Evolution of Crude Oil Quality

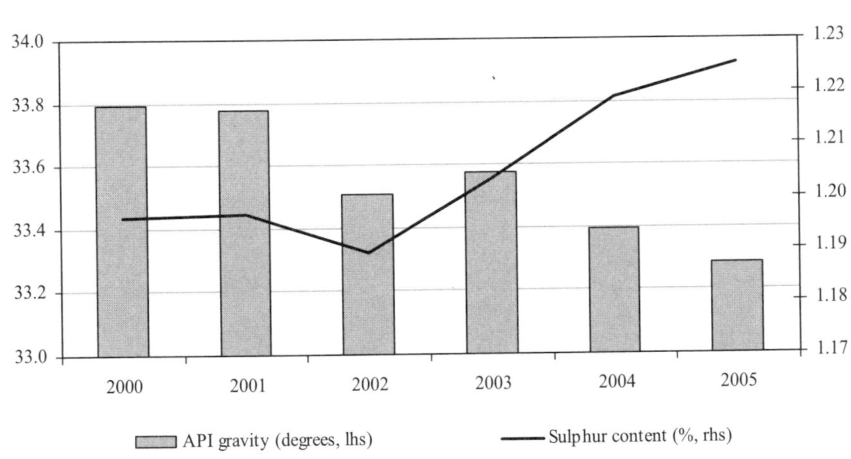

Source: Wood Mackenzie and DB Global Markets Research. Compiled by Repsol YPF Research Department.

Future Crude Oil Quality: Medium and Long Term Forecasts

The deterioration of the crude oil production slate over the last five years was not anticipated by forecasters and it is remarkable that, although future crude quality is a key issue for the oil industry, especially for refining, there are not many forecasts of the future trends in crude oil quality. The following section discusses the medium and long term projections of crude oil quality made by private consultants and public agencies.

1-Private Analysts' Medium Term Petroleum Quality Forecasts

In this section, the medium term forecasts of three well-known consultants are examined: Petroleum Industry Research Associates (PIRA), PetroFinance (PFC) and Cambridge Energy Research Associates (CERA).

PIRA forecasts a changing trend in the quality of produced crude oil towards light and sweet in the period 2005–2010 compared to the period

2002–2005. On the other hand, in 2010 according to PIRA,[7] around 64 percent of total crude oil would be medium and heavy (almost 51 percent medium, 36 percent light and 13 percent heavy) and 60 percent would be of the sour variety (see Figure 4.6).

Figure 4.6
PIRA Petroleum Quality Projections

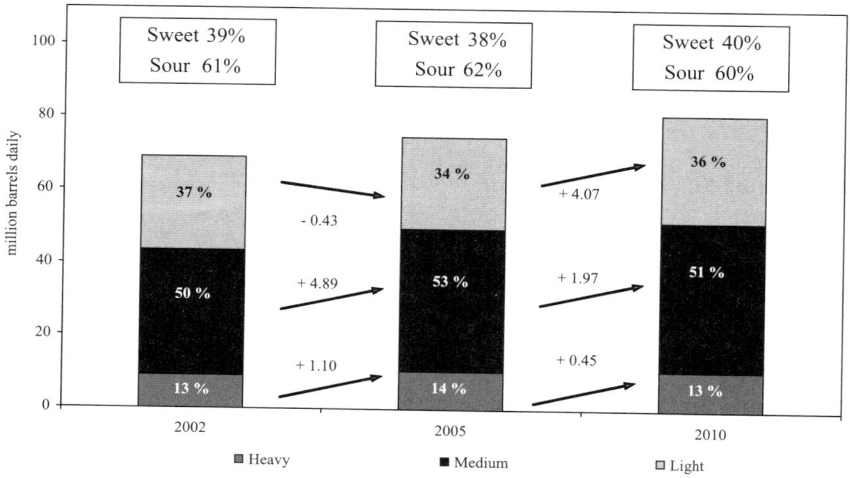

Source: PIRA. Compiled by Repsol YPF Research Department.

However, taking into account the period 2002–2010, these projections represent for 2010 a higher share of medium and heavy crude oil in global production (1 percentage point more than in 2002) and a lower share of light crude (1 percentage point less). According to PIRA, light crude oil production is expected to grow less than medium and heavy (3.64 million barrels daily against 8.41 million barrels) by 2010 from 2002.

This trend is explained by developments during 2002–2005, when medium and heavy crude oil production increased by almost 6 million barrels per day whilst light crude oil production decreased by 430,000 barrels daily. However, for the period 2005–2010 it is expected that light crude oil production will grow more than heavy and medium together (more than 4 million barrels daily as against 2.42 million barrels).

Analyzing the sulfur content projections, PIRA expects that 60 percent of the crude oil produced in 2010 will be sour and the remaining 40 percent will be sweet. These forecasts represent a reduction of sour crude oils production in favor of sweet crude oils in the period 2005–2010. When compared to the year 2002, the share of sweet crude in total production grew by 1 percentage point.

In 2015, according to PFC forecasts (see Figure 4.7), 66 percent of the total crude oil produced will be medium and heavy (52 percent medium, 34 percent light and 14 percent heavy) whilst 61 percent will be sour.

Taking 2002 as the base year in PFC projections (production in 2002 constituted 55 percent medium, 30 percent light and 15 percent heavy), the 2015 figures imply a lower share of medium and heavy crude oil in global 2015 production (4 percentage points less than in 2002) and a higher share of light crude oil (4 percentage points more).

Considering the period 2002–2009, light crude oil production is expected to increase by 3.41 million barrels daily, less than medium and heavy (an increase of 6 million barrels per day). In contrast, in the period 2009–2015 light crude oil production is expected to grow nearly 5 million barrels daily, almost three times more than heavy and medium together (1.74 million barrels per day). The daily growth in the 2002–2015 period is 8.34 million barrels per day of light oil and 7.74 million barrels of medium and heavy.

Regarding sulfur content projections, PFC expects that 61 percent of the crude oil produced in 2015 will be sour and the remaining 39 percent will be sweet. These forecasts imply an increase in the proportion of sour crude oil production during the period 2009–2015.

Figure 4.7
PFC Petroleum Quality Projections

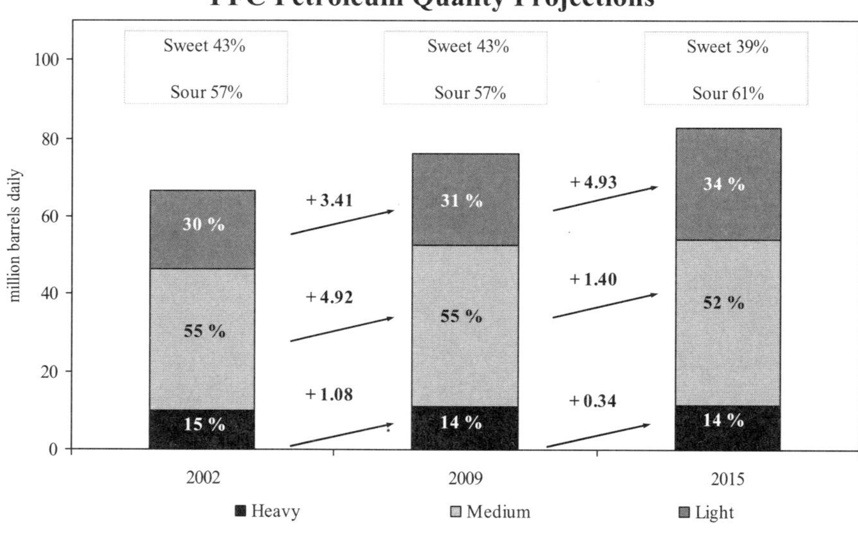

Source: PFC. Compiled by Repsol YPF Research Department.

Regarding CERA forecasts (see Figure 4.8), it can be highlighted that in line with PIRA and PFC projections, CERA expects that in 2010, the medium and heavy crude oil barrels will be higher in number than the light barrels, constituting 64 percent of the total crude supply (with 9 percent heavy, 55 percent medium and 36 percent light).

Nevertheless, taking into account CERA's base year (production in 2004: 59 percent medium, 34 percent light and 7 percent heavy), these projections represent a lower share of medium and heavy crude oil in global 2010 production (2 percentage points lower than in 2004) and an increasing share of light crude oil (2 percentage points higher), similar to PFC projections.

Considering the period 2004–2010, medium and heavy crude oil is expected to grow more than light (8.87 million barrels daily against 7.64 million). This differs from PIRA and PFC views but it must be remembered that the base year for CERA is 2004 and not 2002. Regarding crude oil sulfur content, CERA does not make any explicit forecast.

Figure 4.8
CERA Petroleum Quality Projections

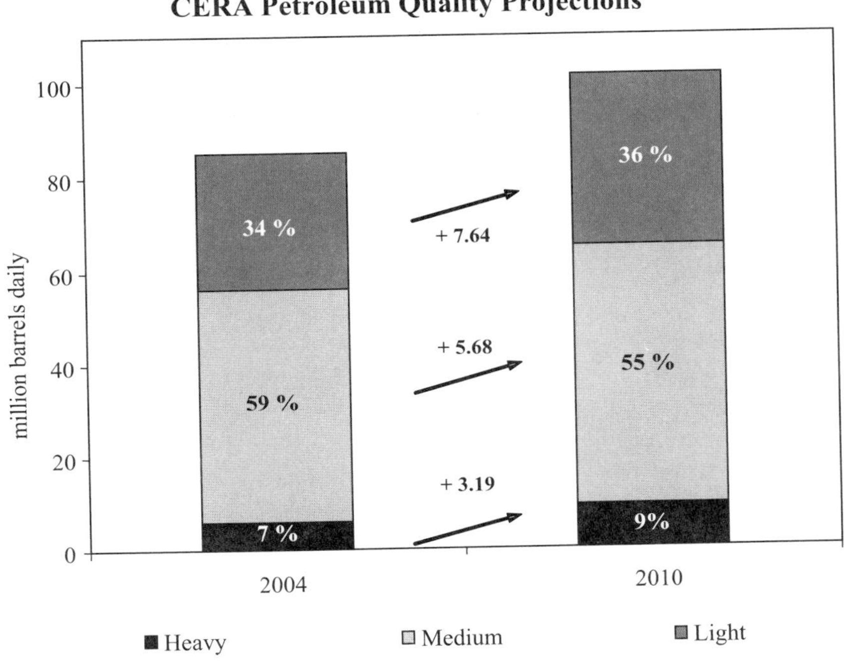

Source: CERA. Compiled by Repsol YPF Research Department.

For 2010, according to Wood Mackenzie, an average crude oil API gravity of 33.37 °API and a sulfur content of 1.23 percent can be expected (see Figure 4.9). These values represent, from the year 2000, a decrease of 0.42° in average API gravity and an increase of 0.4 percentage points in sulfur content compared to the base year 2002.

However, in line with other analysts, from 2006 onwards, a slight improvement in API gravity is expected. This is not the case for sulfur content, for a future increase is expected during the period 2006–2010.

To sum up, according to private analysts, in the medium term, the future trend of crude oil quality is clear: the crude oil produced will be mainly medium and heavy (around 64 percent in 2010), and sour (around 60 percent during the period 2010–2015).

Figure 4.9

Past and Future Trends in Crude Oil Quality Availability

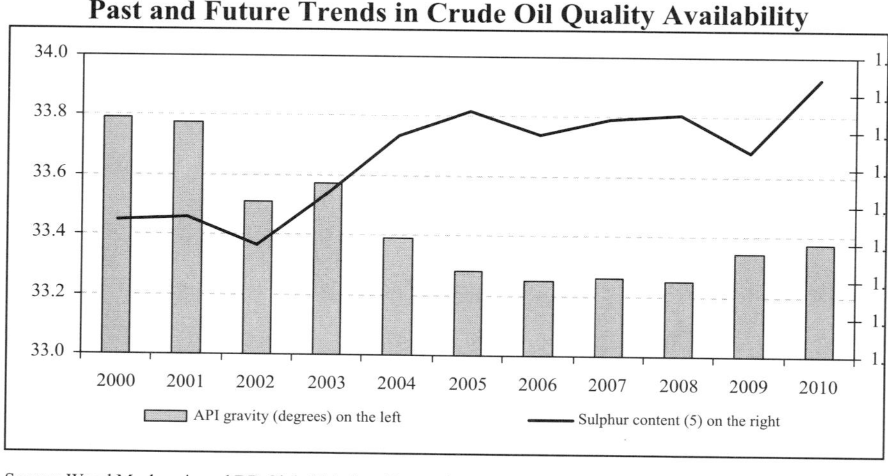

Source: Wood Mackenzie and DB Global Markets Research. Compiled by Repsol YPF Research Department.

Nevertheless, there are differing opinions on the growth rate of medium and heavy crude oil production. Two of the three analysts mentioned above expect that this will decrease in the coming years while the growth rate of light oil production will increase.

2-Public Agencies: Long Term Crude Oil Quality Forecasts

The International Energy Agency (IEA) is the only public agency making explicit projections on future crude oil quality.

According to the IEA[8] (see Figure 4.10), crude oil quality will fall in the coming years and on average, it will be increasingly medium (about 30 °API in 2030). It is important to note that the IEA does not aggregate NGLs to the petroleum produced, so in comparison with other forecasts, the API gravity should be lower or, in other words, the projected petroleum should be heavier.

In its *World Energy Outlook 2005, Middle East and North Africa Insights*, the IEA has included a brief section projecting the average crude oil quality of the Middle East and North Africa (MENA) region, which has a 35 percent share of world oil production.[9] Except for North America and Europe, the remaining production is located in regions characterized by their medium and heavy crude oil production[10] and these projections help to provide a global perspective on the future trend in crude oil quality.

Figure 4.10

IEA: Expected Average Crude Oil API Gravity

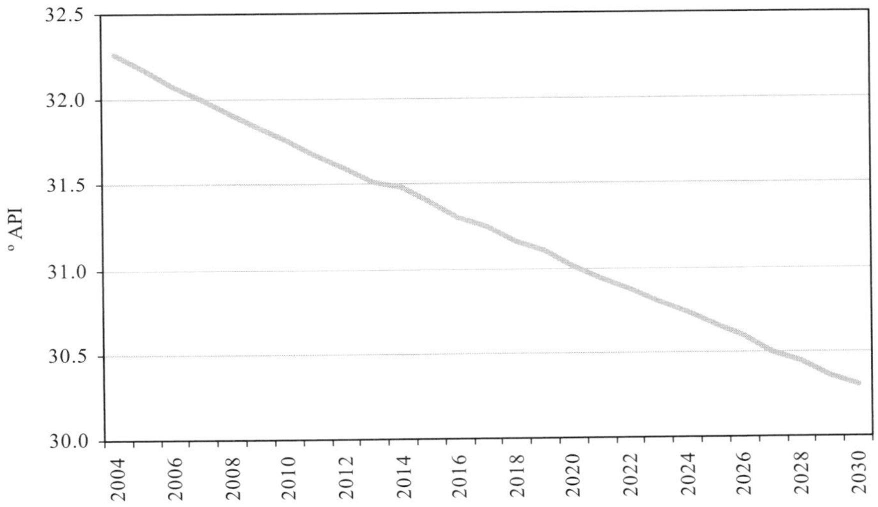

Source: IEA.

According to the IEA (see Figure 4.11), from 2004 to 2030, the average quality of the crude oil produced in MENA is expected to deteriorate although remaining within the medium range. The average MENA crude oil API gravity has averaged 33.2 degrees over the past 35 years and stood at 33.1 degrees in 2004. It is projected to drop to 32.3 °API in 2030, averaging 33.0 °API over 2005–2030.

The decline will be more remarkable after 2015, when more oil is expected to come from the Middle East (where crude oil has lower average API gravity compared to that produced in North Africa). The IEA is also expecting the average sulfur content of MENA crude oil to increase.

Figure 4.11
MENA Average Crude Oil Quality

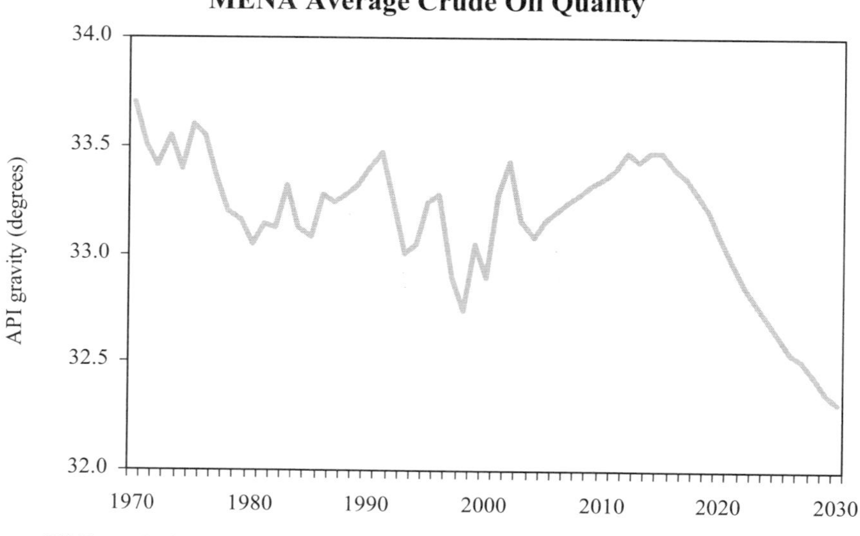

Source: IHS Energy database; *Oil & Gas Journal*; IEA Analysis.

Future Crude Oil Quality According to Expected Production by Region

The implicit projections on future crude oil quality by some analysts and public agencies can be figured out by making a regional analysis of their expected crude oil production forecasts. The forecasts of the Energy Information Administration (EIA), a US government agency will be analyzed in this context.

On the other hand, since the incremental OPEC share in oil supply[11] could be associated with the deterioration of the crude slate quality, any projection of future OPEC and non-OPEC regions could even be taken as an implicit forecast of crude quality. So, future OPEC requirements, according to public agencies, could also be considered as implicit projections of future crude oil quality.

1-EIA Production Forecasts: Assumption of Unchanged Regional Quality

It may be assumed that the regional breakdown of future production projections by the EIA in its *International Energy Outlook*[12] contains "implicit EIA forecasts on future crude oil quality trends." In this context, it is necessary to assume that future quality production by region will not change. It is important to note that the quantities indicated in Figure 4.12 refer to both conventional and non-conventional oil.

As already mentioned, about 75 percent of the current global oil supply comes from regions producing heavy and medium crude oils.

According to the *International Energy Outlook 2006* (see Figure 4.12), most of future crude oil would come from the main oil producing regions. Thus, regions such as the Middle East, the Former Soviet Union and South and Central America, characterized by their medium and heavy crude oil supply, would increase their production about 52 percent, 80 percent and 93 percent respectively from 2003 to 2030.

Another region with a great increase in production is Sub-Saharan Africa, where the light crude share in total production surpasses 60 percent. Although this appears to indicate an improvement in global crude oil quality, there are other regions, such as North America and Europe, with declining production of light crude oils. This would have the effect of balancing the overall situation.

Projections for North America point to a 31 percent increase over 2003 levels, but this additional supply would come from non-conventional crude oils (Canadian oil sands mostly), which would add more than 4 million barrels daily for the forecast period. On the other hand, Europe would cut its production by more than 2 million barrels daily over the period, which would represent a significant reduction of light crude oil on stream.

In conclusion, considering the EIA forecast, and the assumption of no change in region-wise quality, it seems probable that by 2025 the share of medium and heavy crude on total production will increase slightly and therefore no improvement in the average API gravity is expected in the long term.

Figure 4.12
Regional Supply Growth in Key Production Regions (2003–2030)
(Percentages indicate the increase over 2003 production levels)

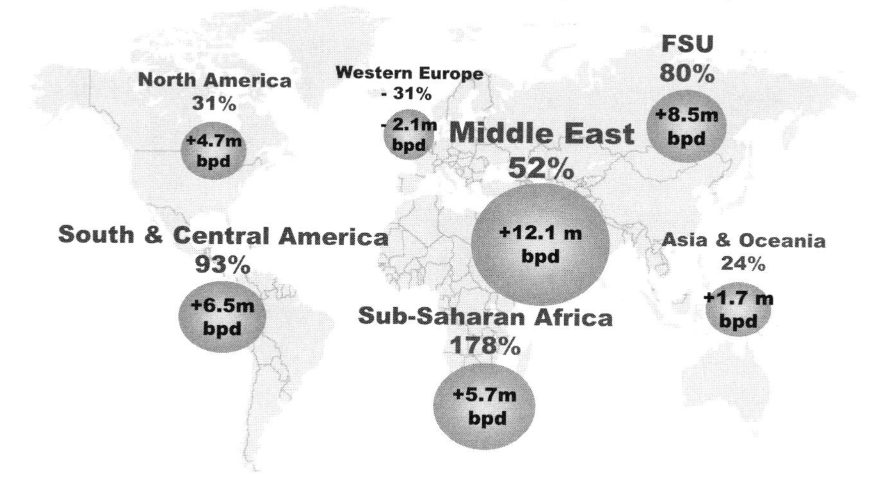

Source: *International Energy Outlook 2006* (EIA). Compiled by Repsol YPF Research Department.

2-Future OPEC Requirements According to Public Agencies

Regarding the projection of future production in OPEC and non-OPEC regions as implicit forecasts of future crude oil quality, the continuous downward revisions of non-OPEC oil production by the public agencies is a cause for concern about future crude oil quality deterioration.

Thus, the EIA forecasts for 2005, 2006 and 2007 of non-OPEC production have been revised downwards by more than 1.6, 1.4 and 0.4 million barrels per day respectively, over the period October 2004–June 2006 (see Figure 4.13). These revisions are particularly pertinent to the subject under discussion because most relate to the OECD areas, where light and sweet crude oils are abundant.

In the same period, the IEA's forecasts of non-OPEC production in 2005 and 2006 have been revised downwards by more than 1.5 and 1.2 million barrels per day, respectively (see Figure 4.14).

Figure 4.13
EIA: Evolution of the Monthly Projections of Non-OPEC Supply
(million barrels per day)

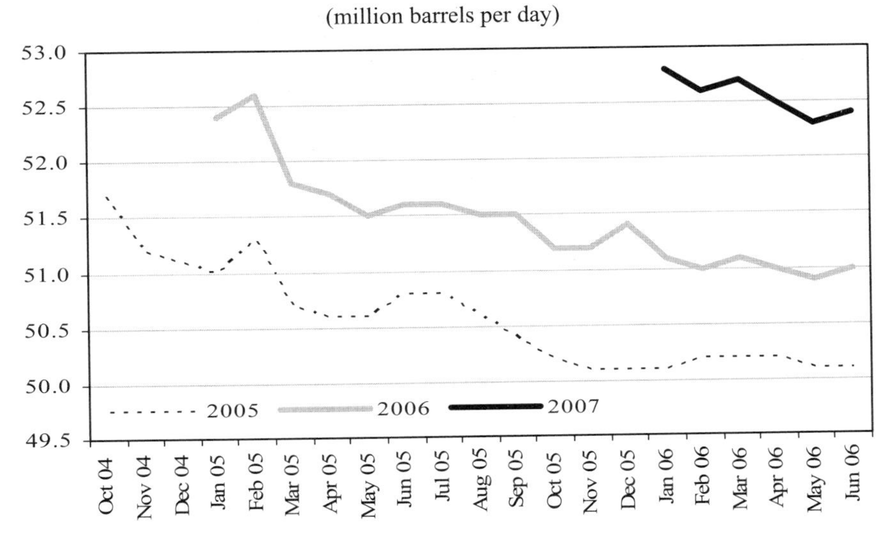

Source: EIA. Compiled by Repsol YPF Research Department.

Figure 4.14
IEA: Evolution of the Monthly Projections of OECD Supply
(million barrels per day)

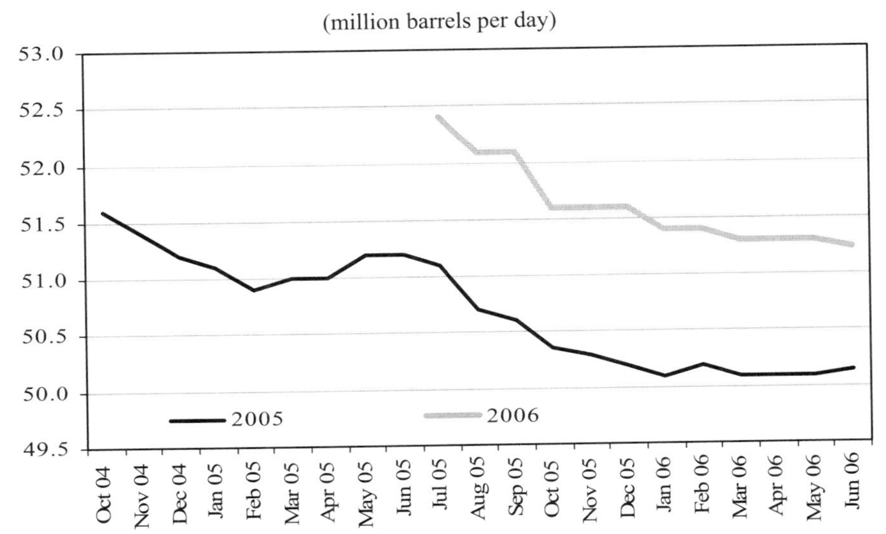

Source: EIA. Compiled by Repsol YPF Research Department.

According to IEA estimates,[13] global oil supply will grow at an annual rate of 1.3 percent between 2004 and 2030, increasing from 82.1 million barrels daily to 115.4 million over the period. This growth will come mostly from OPEC countries, where supply will increase at an average annual growth rate of 2.2 percent, while supply from non-OPEC countries will not grow at all (see Figure 4.15).

Figure 4.15
Growth in Oil Supply

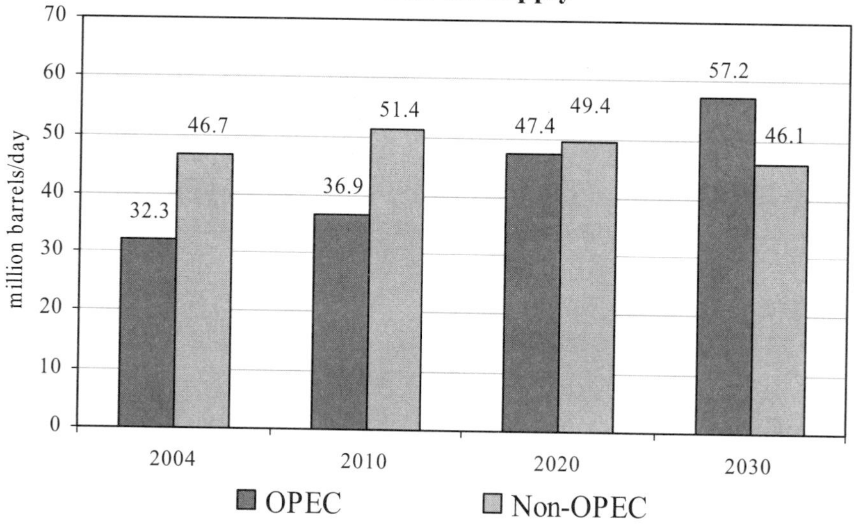

million barrels/day
70
60
50
40
30
20
10
0
32.3
46.7
36.9
51.4
47.4
49.4
57.2
46.1
2004
2010
2020
2030
OPEC
Non-OPEC

Source: *World Energy Outlook 2005* (IEA). Compiled by Repsol YPF Research Department.

The drop in oil supply from non-OPEC countries occurs mainly in OECD countries, particularly North America and Europe. This means that the lowered production of light crude would be replaced by the production of heavier crude oil.

Other public agencies, like the International Monetary Fund (IMF)[14] or OPEC[15] itself, have very similar forecasts. As seen in Figure 4.16, the call on OPEC would increase although it is unclear to what extent, due to present uncertainties about the strength of demand and the rate of depletion in some non-OPEC countries.

However, there is a consensus that if demand continues to grow steadily, the only way to satisfy it would be by supplying heavy and sour crude oils from OPEC.

Figure 4.16

Call on OPEC in 2020: Forecasts by Agencies

million barrels/day
55
50
45
40
35
47.4
38.6
46.2
48.7
IEA
EIA
OPEC
IMF

Source: IEA, EIA, OPEC and IMF. Compiled by Repsol YPF Research Department.

According to these projections more OPEC crude oil production is needed. Therefore, in the short term, the only OPEC crude oil available is the OPEC spare capacity and its quality does not differ from the expected average quality. As seen in Figure 4.17, the oil quality of the OPEC spare production capacity is medium and sour.

Figure 4.17

OPEC Spare Capacity and Crude Oil Quality

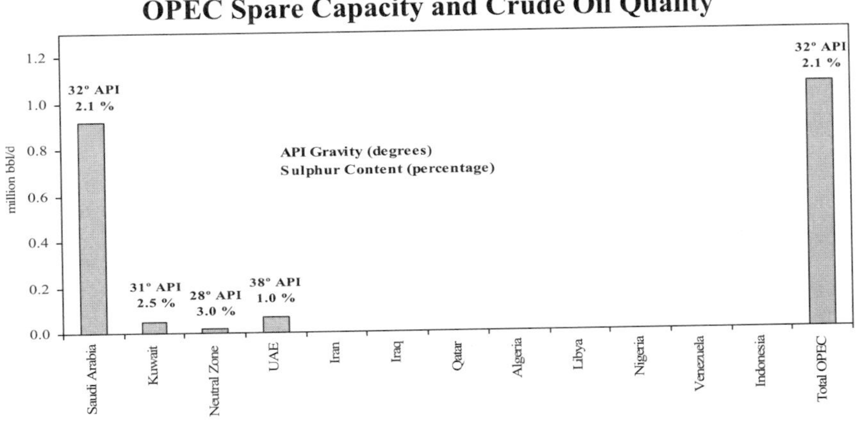

Source: IEA and Energy Intelligence Research. Compiled by Repsol YPF Research Department.

What is more, in the long term, there is always the "law of big numbers." If production has to come mainly from current reserves, what could be expected in terms of future crude oil quality, when 80 percent of total reserves are heavy and sour? (See Figure 4.18).

Figure 4.18

World Proven Reserves by Crude Oil Quality

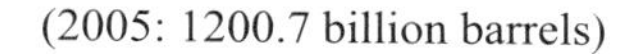

(2005: 1200.7 billion barrels)

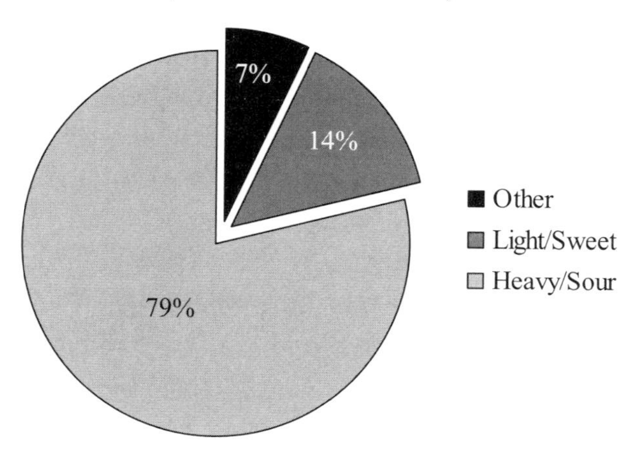

Source: *BP Statistical Review* and Simmons and Company International.

It is true that frontier exploration and prospective estimation of oil to be discovered could change the current picture, particularly if the large surface area open to deepwater exploration is taken into account. However, in the light of current information, future crude oil production will in no case be substantially lighter than in the past.

Now the problem is to ascertain how the forthcoming heavier and sourer crude oil supply can meet future demand requirements, or more particularly, if the final demand requirements will help to solve the imbalance or will serve to amplify the problem.

It is therefore necessary to take a closer look at demand trends.

Demand: Structural Change Leads to Refining Imbalances

On the demand side, the refining sector has to adapt to a completely new environment that is affecting the entire petroleum industry. High global economic growth is driving strong petroleum demand. Nonetheless, final oil consumption is evolving in a different manner depending on the effects of oil price increases on the different final consuming sectors. In this respect, it is the transport sector and its intensive use of clean products that is setting the demand trend.

Additionally, the new environmental regulations and specifications are tightening the products market by reducing the refining yield and consequently creating new challenges for the refining sector. These issues call for further analysis.

Impact of Increasing Transport Sector Growth on Global Oil Demand

According to the IEA, most of the world's current oil demand growth comes from the transport sector. This sector will account for 54 percent of total primary oil energy supply in 2030, compared with the current 47 percent and the 33 percent registered in 1971.[16] Consequently, the transport sector will absorb two-thirds of the increase in total oil consumption.

The weighting of oil in the energy demand generated by the transport sector, in which light products are concentrated, will remain nearly constant at 95 percent for the next 25 years (see Table 4.2). This situation will persist despite the policies and measures implemented in several countries to encourage the use of alternative fuels such as natural gas and biofuels.

According to PIRA[17] all growth in oil demand will be driven by the consumption of clean products, especially those used in the transport sector. Henceforth, the term "clean product" will cover the following petroleum products: gasoline, aviation jet fuels and distillates.

Table 4.2
Global Oil Demand in the Transport Sector According to IEA

	Levels (million barrels per day)				Growth Rates (%)		
	2002	2010	2020	2030	2002-10	2002-20	2002-30
Total Final Petroleum Supplied	77.0	90.4	106.7	121.3	2.0	1.8	1.6
Final Consumption in the Transport Sector	36.4	44.5	55.1	65.4	2.5	2.3	2.1
Final Consumption in the Industry Sector	12.7	14.3	16.4	18.1	1.5	1.4	1.3
Final Consumption in Other Sectors (Agri., Res. & Others)	10.5	11.9	13.7	15.4	1.5	1.4	1.5
Final Consumption in Power and Heat Generation	6.0	8.2	6.4	5.9	0.1	0.3	-0.1

Source: IEA. Compiled by Repsol YPF Research Department.

Figure 4.19
Growth in Demand for Clean Products

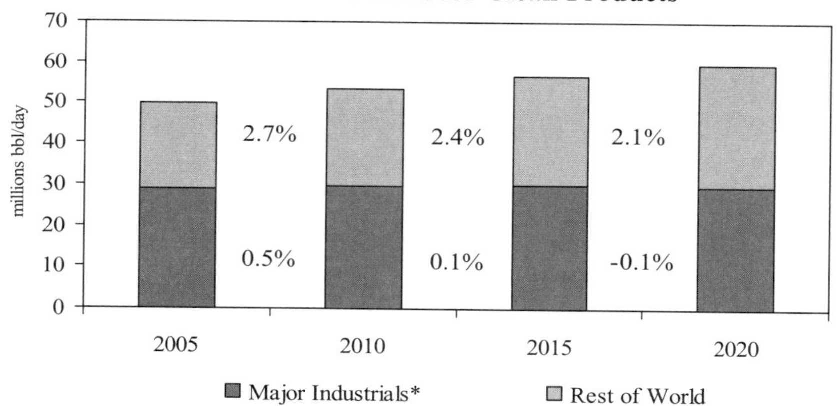

* Major Industrials include Australia, Canada, Japan, New Zealand, United States and OECD Europe (excluding Hungary).
Source: PIRA. Compiled by Repsol YPF Research Department.

As Figure 4.19 indicates and according to most sources, the demand for clean products should continue to grow at a strong pace. Although from 2010, the rate of growth for these products is expected to drop from

0.1 percent to -0.1 percent per annum in industrialized countries such as the United States, Western Europe and Japan, an annual rise ranging from 2.1 percent to 2.4 percent per year is forecast for the emerging countries.

Figure 4.20

World Clean Oil Product Demand

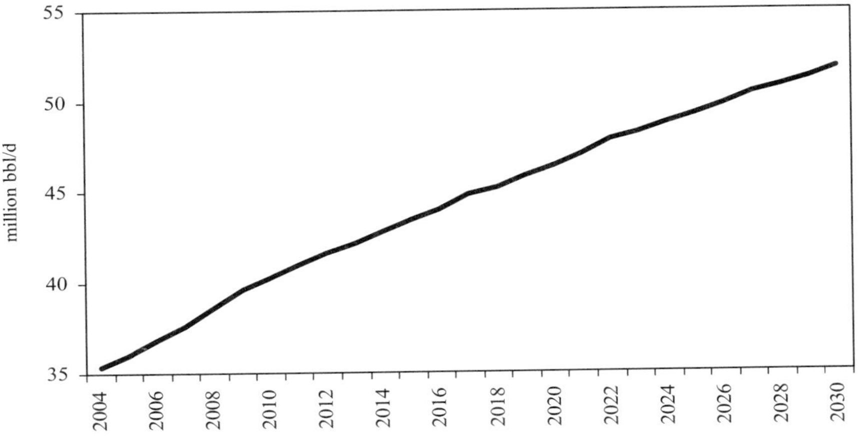

Source: IEA. Compiled by Repsol YPF Research Department.

According to some analysts, the overall demand for clean products will increase almost 1 million barrels per day up to 2015 and, based on IEA estimates, by over 0.6 million barrels per year up to 2030 (see Figure 4.20).

Growth projections for fuel oil, however, indicate that demand for this product will remain flat. In the major industrialized countries demand is expected to fall 2.3 percent per annum between 2005 and 2015, while growth in emerging countries over this period will reach a modest 1.6 percent per annum. Overall, the demand for fuel oil up to the year 2015 will rise less than 0.1 million barrels per day each year (Figure 4.21).

Both medium and long-term forecasts indicate that demand for clean products will account for nearly all consumption growth. Furthermore, looking at consumption indicators in relation to per capita income, it may be inferred that, if consumption in large emerging economies takes the same path as in countries such as Korea during the 1980s, the projections for clean product consumption put forward by public and private agencies will be too conservative.

Figure 4.21
Growth in the Demand for Fuel Oil

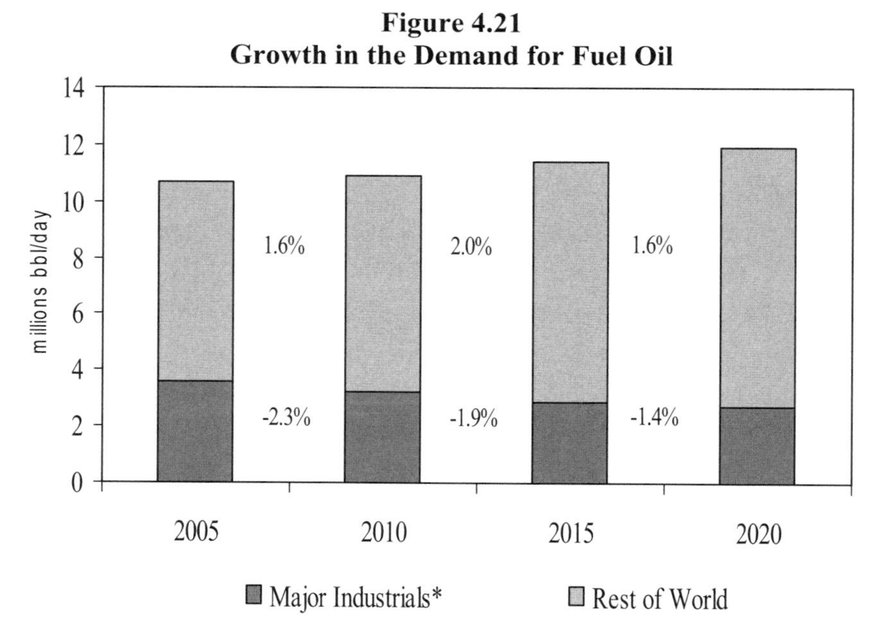

* Major Industrials include Australia, Canada, Japan, New Zealand, United States and OECD Europe (excluding Hungary).
Source: PIRA. Compiled by Repsol YPF Research Department.

In this regard, the growing importance of emerging countries in the world economy is another essential aspect behind the strong oil demand growth, especially in the transport sector.

Drivers of Trend towards Clean Products Demand Growth

1-Acceleration in Global Economic Growth

Economic growth is by far the most important driver of energy demand.[18] In fact, the link between oil demand and economic output has been widely discussed and the main conclusion is that both economic growth and oil demand increase are directly related (see Figure 4.22).

Figure 4.22
GDP and Petroleum Demand Growth

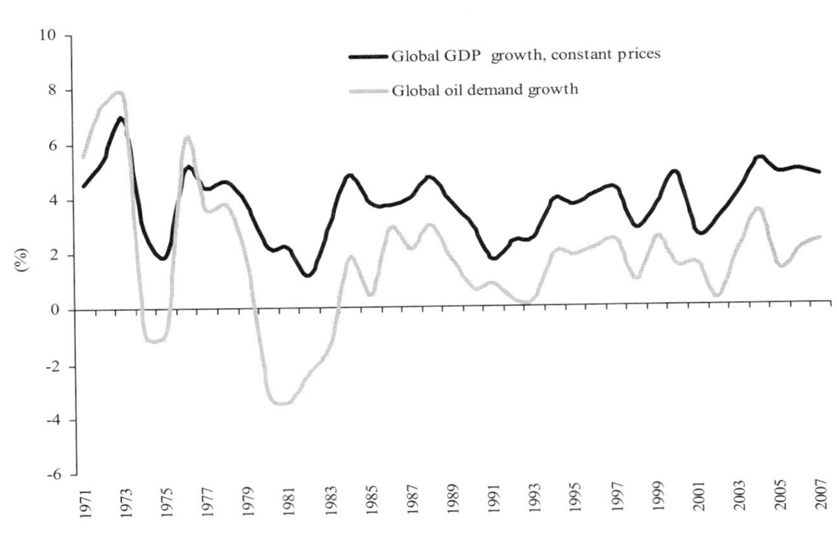

Source: IMF and IEA. Compiled by Repsol YPF Research Department.

Price is another explanatory variable of oil demand, mainly its relative value, in respect of a basket of goods or other primary energy sources.

The IMF's growth forecasts for 2006 and 2007 are 4.9 percent and 4.7 percent, and these figures represent an acceleration over its long-term average global economic growth.[19] According to these strong global economic growth projections, oil demand should stay robust unless high oil prices drive demand down.

Economic growth can be affected indirectly by an oil price hike and its effects on headline inflation. During the period 1970–1990, high oil prices led to an important increase in headline and core inflation. This effect is not currently taking place.

The Central Bank's response to the increase in inflation fueled by oil price rise, particularly when industrialized economies were growing at potential capacity, was one of the reasons for the recessions that followed the oil price increases of the 70s and 80s. As many authors indicate[20] the rise of the real interest rate, which followed these oil crises, worked as a trigger for such recessions.

However, since 2000 the increase in oil prices had different effects on the global economy. Both supply and demand are less affected by the negative effects. To evaluate this, on the supply side, we should consider the lower impact of energy dependency (because of an improvement in efficiency) in the industrialized countries; greater weight of the transport sector; lesser impact on profit margins and the lack of reaction in terms of inflation expectations.

On the demand side, the global impact is lower. There is less probability of a wage-price spiral, interest rate increases by central banks have generally been moderated, the current higher level of taxation on oil products (specially in Europe) implies that international price increases have less inflationary effects than in the 1980s, and oil demand is very resilient because people are not buying less gasoline even when faced with fixed disposable income levels and higher oil prices.

The lower inflationary impact of an increase in oil prices, plus the deflationary effects of globalization in addition to deepening economic integration in recent years has contributed to improving the output-inflation trade-off around the world. As might be seen in Table 4.3, a 10 percent rise in the energy components of the Consumer Price Index (CPI) triggered a smaller estimated impact on core CPI during the period 1990–05 than during 1970–89.[21]

The tamer effect of high oil prices on inflation has avoided a stark and rapid increase in the official interest rate.[22] In fact, real interest has remained historically low, contrasting sharply with past episodes of oil price increases (see Figure 4.23).

Table 4.3
Estimated Impact of 10 Percent Rise in the Energy Component of CPI on Core CPI

	1970–1989	1990–2005
United States	1.7%*	-0.1%
Euro Area	3.2%*	0.3%
Japan	3.1%*	0.5%
United Kingdom	3.4%*	-0.8%
France	3.4%*	-1.2%
Italy	3.9%*	2.8%
Canada	2.6%*	0.0%

Note: Results are based on a four-lag vector auto regression. Statistical significance typically exceeded 1 percent in the 1970–89 period.

Source: Citigroup.

Figure 4.23
Real Interest Rates and the Price of Oil (WTI)

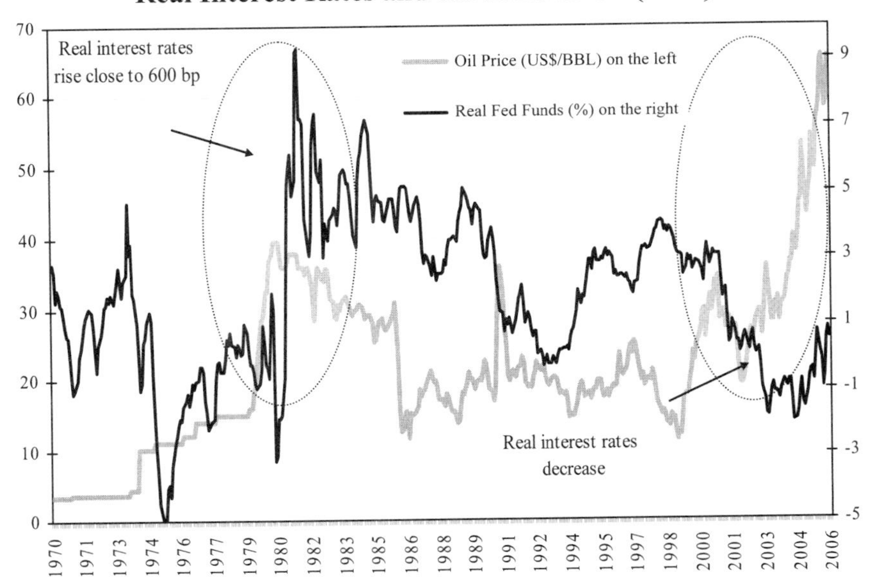

Source: IMF and US Federal Reserve. Compiled by Repsol YPF Research Department.

There is no reason to believe that monetary authorities will respond by raising interest rates again or that real interest rates will rise at a similar pace as in 1973 or 1980. This is mainly because the pass-through effect of oil price increases on core inflation is much lower than 10 years ago.

A robust economic growth forecast, which is not being affected by the high oil prices, confirms the argument that oil demand will continue to grow in the coming years.

2-Greater Importance of Emerging Countries in the World Economy

The dynamism of emerging countries, especially China's growth, has fostered global economic vigor and consequently global oil demand drive.

In fact, China has progressed from being a net oil exporter in 1992 to becoming the world's third largest net importer in 2004, with 3 million barrels per day of imports.

With regard to the global growth estimated by the IMF for 2004 and 2005 (5.3 percent and 4.8 percent, respectively), China would account for 23 percent, Asia (excluding China) would account for 20 percent, and Latin America would represent 10 percent. The implications for oil demand growth are dramatic. The rate of growth in Chinese oil consumption for the last five years has been close to its GDP growth rate.

Furthermore, in emerging Asian markets, especially in China, the elasticity in oil demand with respect to income per capita or GDP is higher than expected and clearly exceeds the long-term estimates put forward and applied by public energy agencies. This greater elasticity is largely the result of a generalized motorization phenomenon, but it is also influenced by the unexpectedly heavy weighting of investment on the part of emerging countries in manufactured goods for export and transport infrastructures.

Figure 4.24
China Oil Demand

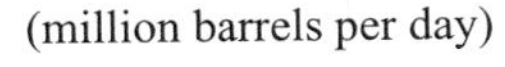
(million barrels per day)

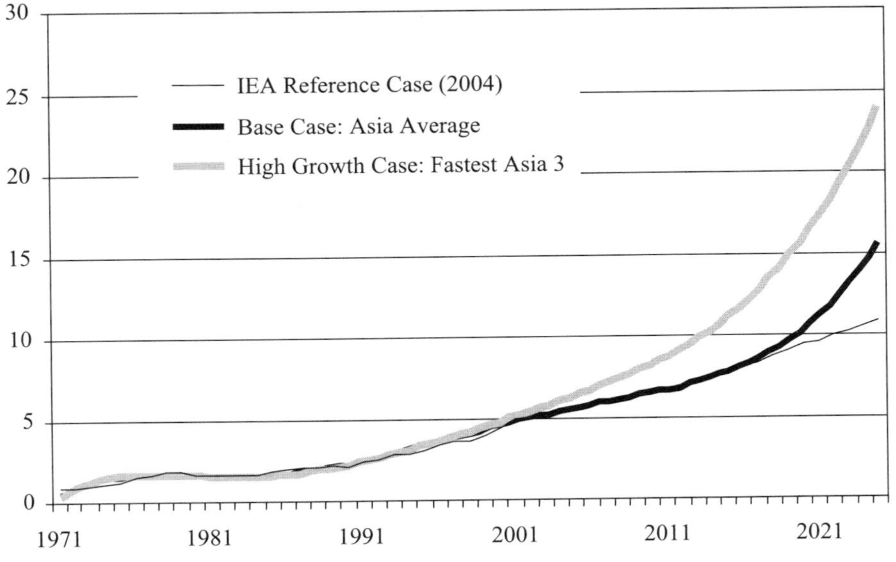

Source: IEA. Compiled by Repsol YPF Research Department.

As seen in Figure 4.24, according to the IEA base case, China would demand 13 million barrels daily by 2030. If China behaves as average Asian countries did, its demand would be near 15 million barrels. However, if the transition is similar to that of the fastest Asian countries, its demand would reach 22 million barrels daily by 2030.

3-Substantial Increase in the Number of Vehicles

Growth in fuel requirements for the transport sector is particularly relevant when analyzing trends in product demand. Oil product demand for transportation is especially strong in developing countries, in line with income growth and infrastructure development. Hence, the fleet of passenger vehicles in China, the world's fastest growing market for new cars, has grown more than 9 percent per annum over the last five years, compared to a 3 percent annual growth worldwide.

This market has an enormous growth potential: the penetration rate in China is 16 vehicles per 1,000 inhabitants, compared with 812 in the United States and over 500 in Europe (see Figure 4.25). The vehicle fleet in other Asian countries, including Indonesia and India, is also growing rapidly. According to several estimates, 20 years from now, China could have more cars than the United States while India could reach the same situation in 30 years time.

Figure 4.25
Vehicle Ownership Projections (2002–2030)

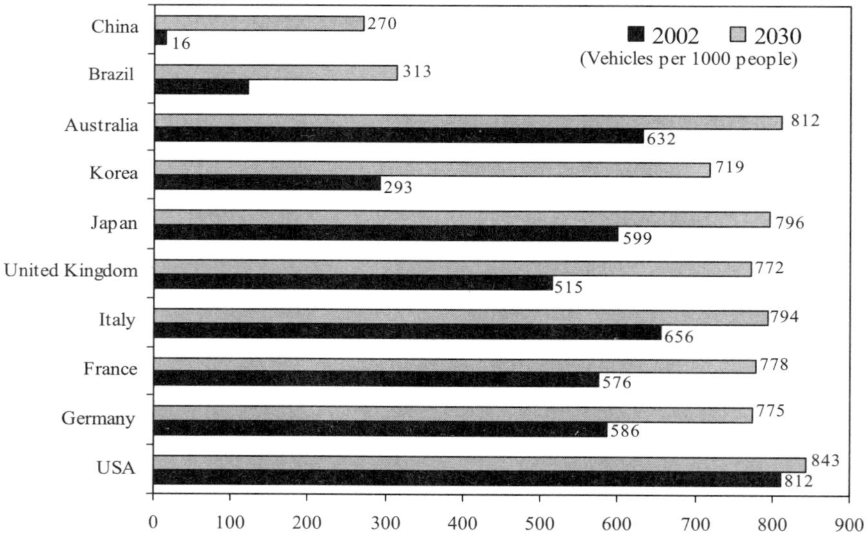

Source: United Nations Annual Report. Compiled by Repsol YPF Research Department.

Looking closely at the experience of countries that have evolved from low to medium/high income in the past 25 years, it may be noted that growth in oil demand for the transport sector increased at the same rate as per capita income. In fact, OECD countries started to experience rapid vehicle ownership growth when they achieved levels of about US$2500 of per capita income in PPP terms.

Figure 4.26
Vehicles Ownership vs. GDP

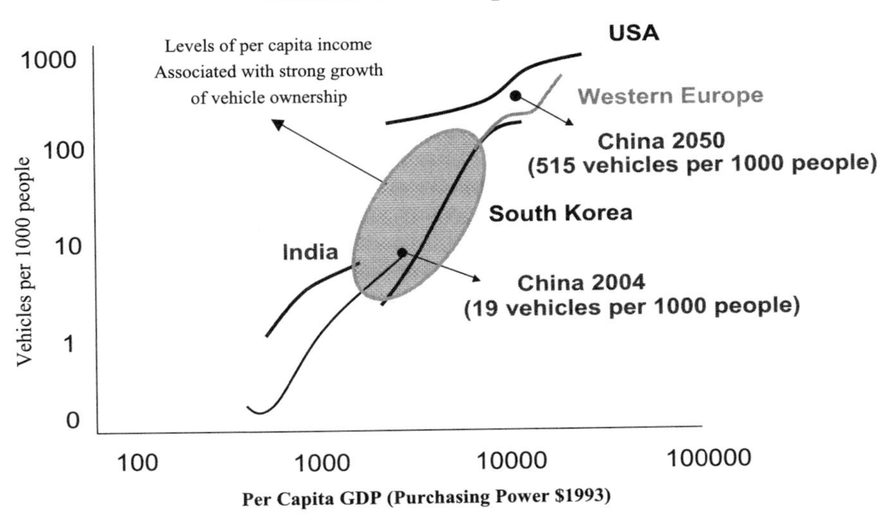

Source: United Nations Annual Report and IMF. Compiled by Repsol YPF Research Department.

Accordingly, the number of cars will increase substantially in emerging countries as the number of people entering the $1500–2500 range of per capita income increase. To fully appreciate the potential growth in demand for transport, suffice it to say that over 80 million people per year in developing countries are expected to reach a per capita income of 3000 dollars in the next three years.

4-Resilience of OECD Demand

Another key factor behind the rise in oil consumption is the resilience of demand, which in turn, is related to the increasing participation of the transport sector in the final consumption of oil, as opposed to other sectors or destinations, such as the residential and industrial sectors.

In the OECD countries, oil demand in the transportation sector represented 70 percent of total demand in 2004 against 49 percent in 1979. The lack of an alternative car fuel makes oil demand in this sector very resilient to oil price increases.

Demand in industrialized countries does not suffer in a price scenario of around 40 dollars per barrel because, in contrast to what happened in the seventies and eighties, petroleum is mainly used in the transport sector and essentially in the private utilization of cars.

Demand resilience is also linked to the profile of car buyers in the OECD region and the kind of car in demand.

Part of the incremental demand is linked to the fact that big cars (sport utility vehicles or SUVs) are considered a luxury good, which medium and high income people are demanding increasingly, while low income citizens also tend to follow that pattern, whenever possible. This is evident in the increasing share of SUVs in the American car market (Figure 4.27). Additionally, one has to point out that a similar pattern is increasingly visible in Europe.

Figure 4.27
Sport Utility Vehicles: US Market Share Percent

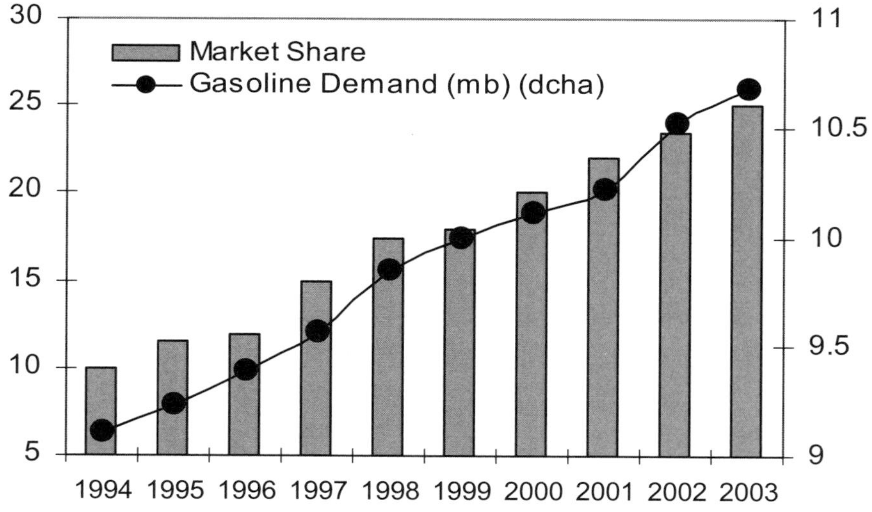

Source: General Motor Cars. Compiled by Repsol YPF Research Department.

So, although clean product demand was expected to diminish in the industrialized countries in a scenario of high prices, the rising use of

SUVs in the OECD is sustaining a strong demand for clean fuels, offsetting the expected trend.

On the other hand, this boom in the purchase of SUVs has not been slowed down by high fuel prices. According to Merrill Lynch Commodity Research, on average, a 10 percent increase in world petroleum product prices would only slow down the rate of global oil demand growth by 0.5 percent.[23] Price elasticity of oil demand varies considerably by region, and some parts of the emerging world can be particularly sensitive to oil price increases (see Figure 4.28).

Figure 4.28
Price Elasticity of Oil Demand across the World

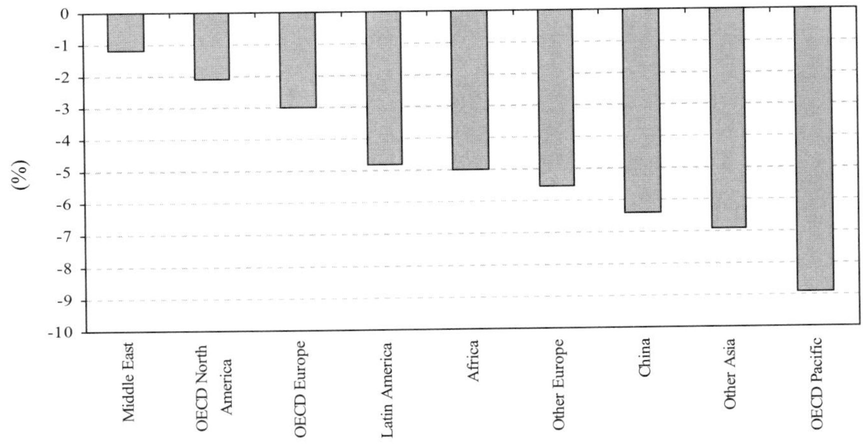

(%)
0
-1
-2
-3
-4
-5
-6
-7
-8
-9
-10
Middle East
OECD North America
OECD Europe
Latin America
Africa
Other Europe
China
Other Asia
OECD Pacific

Source: Merrill Lynch Commodity Research.

The one market where price elasticity of demand has proved to be exceptionally low is the United States. According to Merrill Lynch, there is a close inverse relationship between the US personal savings rate and gasoline prices, suggesting that US consumers are growing increasingly reliant on debt to pay for higher fuel cost. Gasoline consumption has increased as a share of disposable income, but consumers have continued to borrow rather than cut back on other expenditure (See Figure 4.29).

Figure 4.29
US Gasoline Consumption
(percent disposable real income)

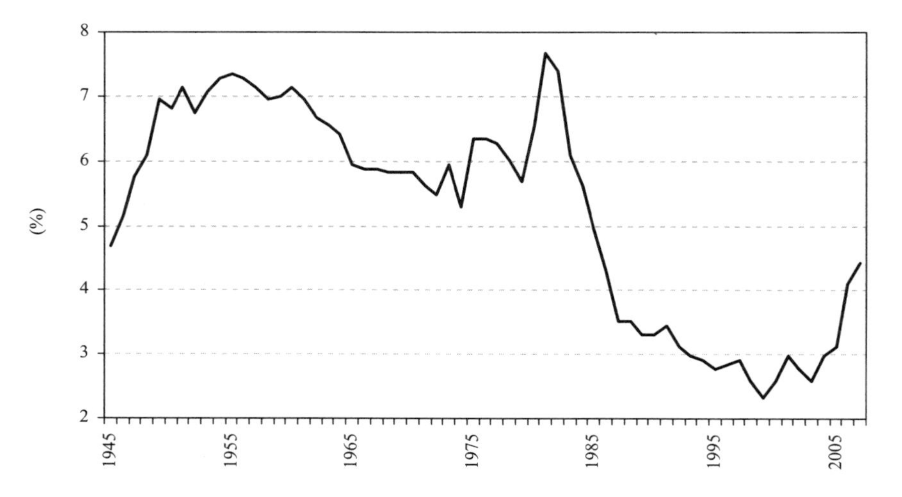

Source: Merrill Lynch Commodity Research.

Structural Change: Green Rules Spur Product Specifications

Policies on global warming have experienced important changes over the past decade. In addition to implementing operating regulations, such as the treatment of wastewater or gas emissions, the refining industry has also had to adapt the structure of its processes to ensure that its end products meet the most stringent product specifications, and will continue to do so in the future.

Initially, the scope of these specifications was mainly technical: the octane content in gasoline, smoke point for kerosene, or fuel viscosity. Now, however, environmental protection stipulations include many more requirements: for example, in gasoline, limiting the content of sulfur (when burnt, this compound produces sulfur dioxide, which oxidizes when released into the air and produces acid rain in contact with water), benzene (a carcinogenic hydrocarbon), olefin and aromatics (hydrocarbon families that are highly reactive in the air we breathe and contribute to the formation of the ozone in the troposphere where high ozone concentrations are toxic); vapor pressure (which must be limited in order to reduce emissions from vehicles and at service stations), and even call for the presence of oxygenated compounds not naturally found in petroleum in order to enhance full combustion within the engine.

The schedule for implementing new sulfur content specifications in different countries is shown in Table 4.4. It may be seen that the most stringent specifications are applied in European Union countries, where, in less than 7 years, sulfur content in fuel must be reduced by approximately 98 percent. This trend is increasingly global. In 2006, 70 percent of gasoline and 50 percent of diesel world demand will be affected by new low sulfur standards. According to the specification requirements, the United States, China and India will also have to make a big effort in the next few years to reduce their sulfur emissions.

Table 4.4

Sulfur Content Specifications as per Implementation Schedule

	Before 2004	2004	2005	2006	2009	2010
European Union						
Gasoline	500	150	50		10	
Diesel	500	350	50		10	
United States						
Gasoline	300	120	90	30		
Diesel	500	500		15 (1)		
China						
Gasoline	800	500	150 (2)			150 (3)
Diesel	2000	500	350 (2)			350 (3)
India						
Gasoline	1000	500	150 (4)			150
Diesel	2500	500	350 (4)			350

(1) In June 2006, 80% of the diesel must reach 15 ppm, in 2010 it must reach 100%. (2) Implemented in Beijing before the Olympic Games of 2008. (3) By obligation to all nations. (4) Implemented in the biggest cities in 2005, by obligation.

Source: PIRA. Compiled by Repsol YPF Research Department.

This modification in petroleum product specifications is actually changing the nature of demand. There is a global timetable in place from now until 2010 for implementing these new requirements, aimed at reducing sulfur content in practically all parts of the world, which will require strong initial investments to adapt the equipment and units involved.

This process of regulatory change has a decisive impact on the refining sector. First, these regulations tilt demand towards fuels with a lower sulfur content, which should mean higher refining margins for sour crude oil. These margins will have to be higher for a prolonged period in order to allow an adequate return on the investments necessitated by such new specifications. Second, in certain countries, manifold specifications place a burden on the logistics chain, making the end product more expensive.

Last but not the least, without variation in the other conditions, the new specifications imply a reduction of existing and idle refining capacity, for two reasons. First, because refineries failing to implement the new systems will no longer be able to produce legally commercialized products. Second, because the process of desulfurization implies an increase in the auto consumption of energy needed for hydrotreating thus resulting in a reduction in refinery yields.

It can be argued that the new specifications will translate into higher margins because they speed up the destruction of idle capacity. The effective shrinking of the system's capacity is aggravated by other environmental regulations and the moderate levels of investment in hydrogen. The reason being that investment in desulfurization has historically produced a lower return than in other processes, such as cracking or coking.

The Refining Sector: When Supply Meets Demand

In the previous sections, supply and demand drivers have been separately analyzed. Now it is necessary to put together both supply and demand considerations. For supply to meet demand it is necessary to have a good and sound refining industry, especially when the main conclusion of the previous section is that there is an increasing mismatch between final demand requirements and the quality of supply (Figure 4.30).

Figure 4.30
IEA Quality Supply and Demand Trends Forecast

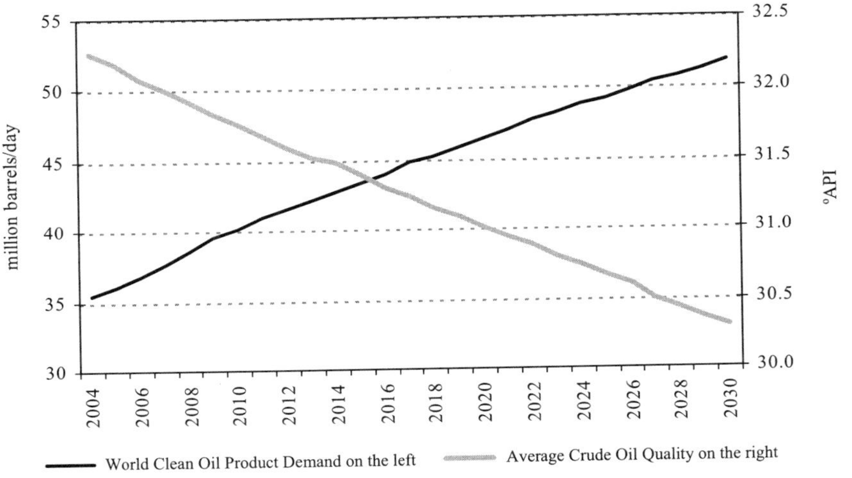

Source: International Energy Agency (IEA). Compiled by Repsol YPF Research Department.

This imbalance can only be solved by greater investment in refining. This section begins by defining the concepts of refining margins and product price differentials. These differentials, along with crude oil price spreads, are fundamental in determining the evolution of refinery margins. An explanation of these differentials is needed to assess the prospect for refining margins.

The second part of this section pinpoints the challenges of "primary supply considerations" (crude quality availability) and "secondary supply considerations" (the capacity of the refining industry to deal with heavier crude oils and lighter demand and respond to the more restrictive environment regulations concerning sulfur content). In this respect, the present bottlenecks in the refining industry and the expected investment in the sector are studied. Finally, a review is undertaken of the factors for and against investment in the refining sector.

Definitions and Relations between High Margins and Price Differentials

Before analyzing the refining sector situation, it is necessary to provide background information on terms and concepts used in refining that will appear throughout this chapter.

1-Refining Margins: Definition and Historical Evolution

The refining margin per barrel of processed crude oil is a profitability indicator applied to the operation of any refinery. This figure, generally expressed in dollars per barrel, can be properly defined as the revenues obtained on production, minus variable operating and raw material costs, all divided by the distillation volume.

Nevertheless, there are many refining margins or refining margin indicators. Most investment banks follow standard or benchmark refining margin indicators based on the so-called 3:2:1 or the 5:2:2:1 formulas, calculated on the basis of the difference between product and crude oil prices in a specific region in turn.

The 3:2:1 formula (2 units of gasoline, plus 1 unit of heating oil, minus 3 units of crude oil) gives a refining margin whose product yield is greater in gasoline than in medium products, whereas the 5:2:2:1 formula (2 units of gasoline, plus 2 units of heating oil, plus 1 unit of fuel oil, minus 5 units of crude oil) gives a refining margin whose product yield is greater in heavier products than in gasoline.

Some of the most extended margin indicators, and those that are used throughout this chapter (see Table 4.5), are the ones defined by the IEA for cracking operations (Rotterdam, Mediterranean, Singapore and US Gulf Coast margins) and for coking (US Gulf Maya margin).[24]

Table 4.5
Product Yields Used for Refinery Margin Calculations (Percent)

	North West Europe Brent Catalytic Cracking	Mediterranean Urals Catalytic Cracking	Singapore Dubai Hydrocracking	US Gulf Coast WTI Cracking	US Gulf Coast Maya Coking
Petroleum Gases	3.1	2.6	5.8	–	2.3
Total Mogas & Naphtha	37.7	31.4	24.7	51.8	62.2
Total Distillate	41.8	39.1	50.0	32.7	26.7
Total Fuel Oil	12.8	21.5	22.3	13.5	1.5
Total	**95.4**	**94.6**	**102.8**	**97.96**	**92.7**

Source: International Energy Agency (IEA). Compiled by Repsol YPF Research Department

Refining margin performance has changed significantly in recent years. Following a moderate decade, refining margins started a sharp increase in 1999 only to drop back in 2002. Economic weakness worldwide was the main reason for the cutback in refining margins that year.

Thus, the low refining margins in Europe, the United States and Japan in 2002 (see Figure 4.31) resulted from a rise in feedstock prices driven by growth in emerging countries, which was not passed on to refined product prices because of sluggish final demand in OECD countries. The negative impact of the 9/11 terrorist attacks on margins, when lower demand led to an accumulation of oil stocks, should not be forgotten.

Figure 4.31
Historical Performance of Monthly Refining Margins

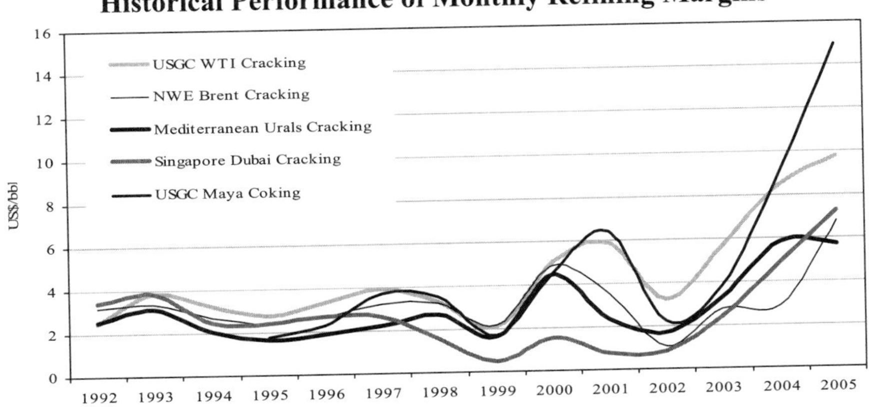

Source: IEA and *Oil & Gas Journal*. Compiled by Repsol YPF Research Department.

There was a clear upturn in margins during 2003, but it was not until 2004 that margins and price differentials between light and heavy crude oils, as well as the differentials between clean (gasoline, kerosene, gas oil) and dirty (heavy fuel oil) products, reached record highs. The 2004 record margins were even surpassed by the levels achieved in 2005.

2-Product Price Spreads and Crude Price Differentials

According to the previous definition of refining margins, it is evident that the wider the price differential between final products and crude oil, the higher the refining margin. This, however, is not the only relevant differential. Analyzing other differentials such as clean versus dirty products or differentials in product price and crude oil price spreads can help to evaluate the evolution and outlook for refining margins at each refinery center, and can even help to assess the weight that supply and demand exercise on margin performance.

For example, it could be argued that when price differentials between heavy and light crude oil narrow or widen in a more noticeable way than product prices spreads do, margins are more influenced by supply considerations. Conversely, when product price spreads widen or narrow more than crude oil price differentials, margins are largely influenced by changes in demand conditions. In fact, the problem is more complex, for there is a close correlation between both differentials and it is necessary to determine which of them is really responsible for the variation in the other.

PRICE DIFFERENTIALS BETWEEN LIGHT AND HEAVY CRUDE OILS

The differential between light and heavy crude oil prices is an indicator of the behavior of refining margin performance. When the spread becomes wider, complex refineries benefit from the lower cost of heavy crude oil and obtain higher margins through conversion. The relationships are described in Figures 4.32, 4.33 and 4.34, summarizing this mechanism.

Figure 4.32
Urals Crude Refining Margin and Brent/Urals Price Differential

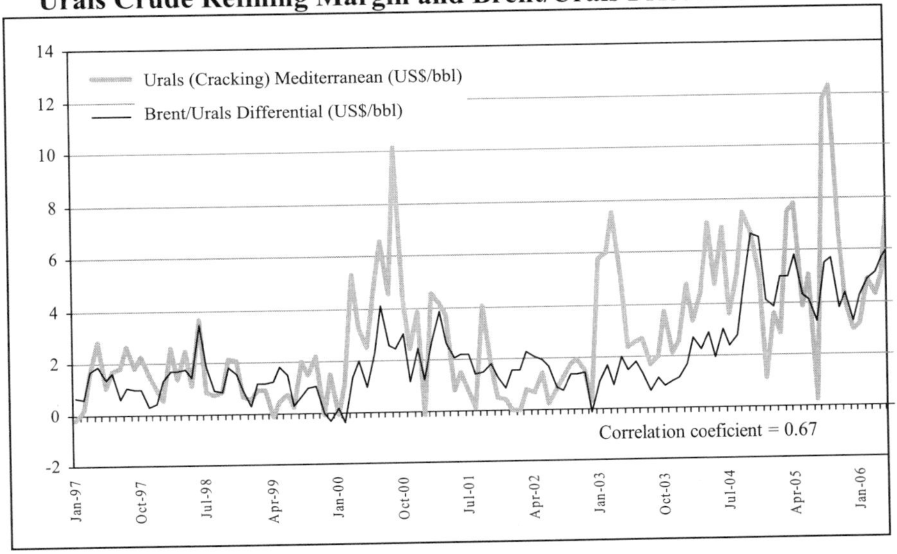

Source: EIA and IEA. Compiled by Repsol YPF Research Department.

Figure 4.33
Dubai Crude Refining Margin and Brent/Dubai Differential

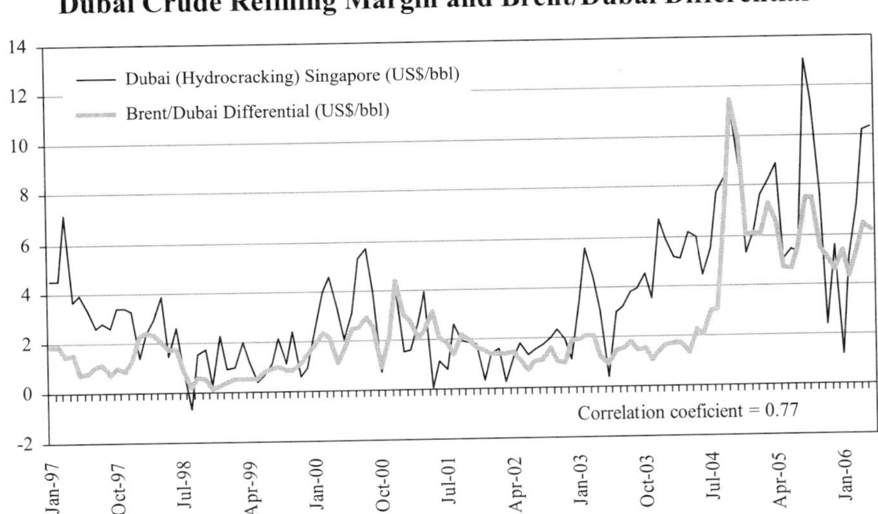

Source: EIA and IEA. Compiled by Repsol YPF Research Department.

Figure 4.34
Relationship between Coking Margin and Crude Oil Price Differential

Maya (Coking) USGC (US$/bbl) on the left
WTI/Maya Differential (US$/bbl) on the right
Correlation coeficient = 0.85

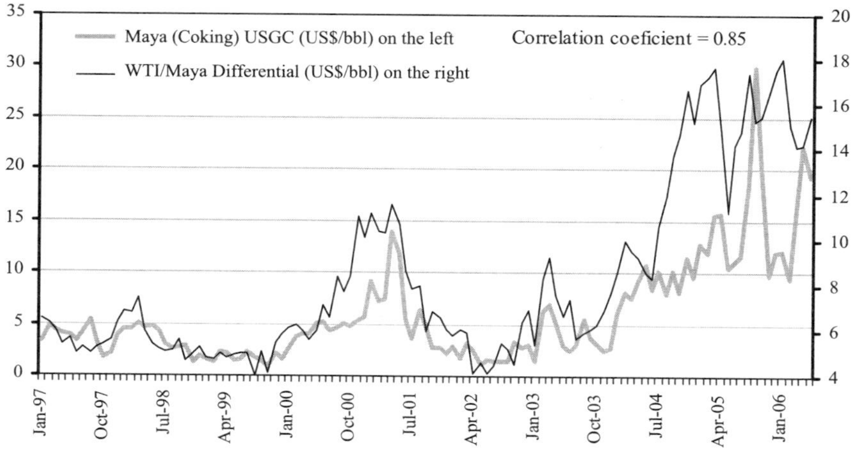

Source: IEA, Bloomberg and Reuters. Compiled by Repsol YPF Research Department.

The relation between crude oil price spreads and refining margins evolution can be analysed by region. For example, studying the relation between the Brent (38° API)/Urals (32.5 °API) spread and Urals cracking margin, results in correlation of almost 0.7.

The case for Dubai (31.2 °API)/Brent spread vis-à-vis Dubai cracking or WTI (38.7 °API)/Maya (21.8 °API) spread vis-à-vis Maya Coking, also results in a high correlation of about 0.8.

In fact there is a causality relation from crude oil price differentials to refining margins, as can be seen in Appendix-I.[25]

CLEAN-DIRTY PRODUCT PRICE DIFFERENTIAL

Another indicator for refining margins is the differential between the price of clean and dirty products. The main input in a complex refinery's conversion unit may be the equivalent of heavy fuel oil, which when processed, makes it possible to obtain clean products with higher added value.

Taking as a reference the motor gasoline (mogas)/fuel oil price differential and the Northwest European (NWE) Brent cracking margin (Figure 4.35), the higher the price differential between clean and dirty products, the wider the associated refining margin. A sustained widening trend is noticeable in motor gasoline/fuel-oil price differentials from 2002 onwards.

Figure 4.35
Brent Oil Refining Margin and Motor Gasoline/Fuel Oil Differential

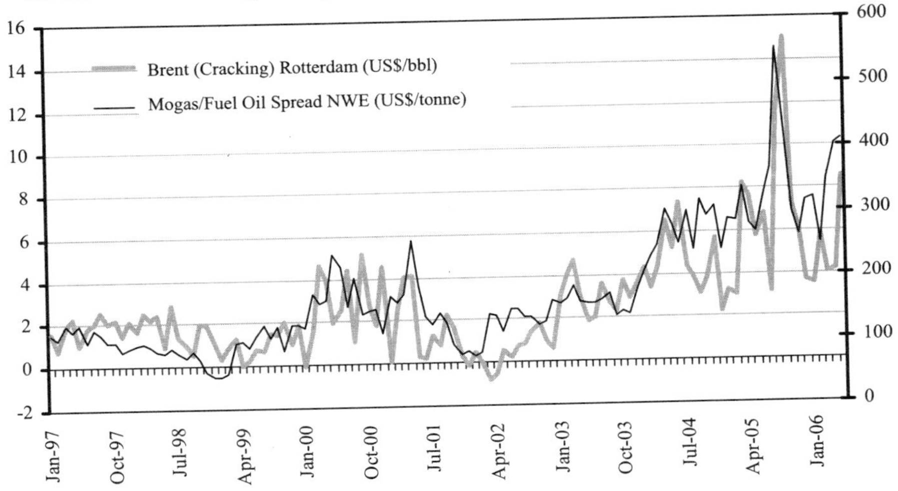

Source: EIA and IEA. Compiled by Repsol YPF Research Department.

The upsurge in margins and price differentials to reach record levels in 2004 and 2005 was accompanied by all-time high crude oil prices. Many of the factors explaining price evolution were also responsible for margin performance. Nevertheless, certain price information is essential for understanding the relative importance of demand and supply factors in refining. It may be argued as follows:

- When margin performance is determined by demand conditions, there is a very close or increasing correlation between the differential for clean/dirty product prices and the refining margin.
- When supply conditions are the determining factor, refining margins bear a tight correlation with the differentials in crude oil prices.

After studying the evolution of price spreads or differentials, it is clear that there is a very high correlation between refining margins, crude differentials and product price spreads in 2004 and 2005. In 2000, however, when margins were also high, the correlation between crude differentials and refining margins diminished considerably, remaining high with respect to the clean/dirty product price spreads. This could suggest that higher margins in 2004 and 2005 were driven by both demand and supply factors, whereas in 2000 they responded more to demand conditions.

Supply considerations are linked to the capability of dealing with heavier crude and, therefore, with the global refining upgrading capacity. However, one has to be cautious when inferring such conclusions, because product price spreads and the absolute price level of crude oil are also highly correlated (Figure 4.36).

Figure 4.36
Relationship between Brent and Product Price Differential (1997–2006)

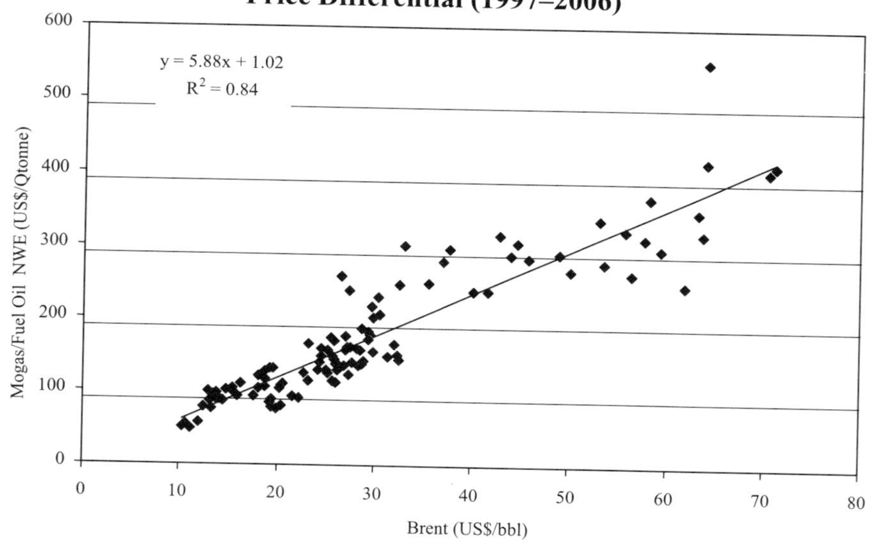

Source: Bloomberg, Reuters and Repsol YPF Research Department.

Current Bottlenecks in the Refining Sector

The evolution of idle refining capacity, defined as distillation potential against oil demand, has clearly diminished over the past two years. This leaves little room for maneuvering if, as forecast earlier in this chapter, demand shows sustained growth in the coming years.

As seen in Figure 4.37, the difference between final demand and distillation capacity is dwindling. However, this potential bottleneck in simple distillation is not the most important or relevant nowadays. Over the last years, the gradual switch of demand towards lighter fuels in the transport sector created a real bottleneck of significant relevance in 2004, attributable to a shortage of additional upgrading capacity in the global refining system. This would be the key factor behind the very wide crude price differentials in 2004.

Figure 4.37

Demand–Supply and Refining Capacity (Distillation)

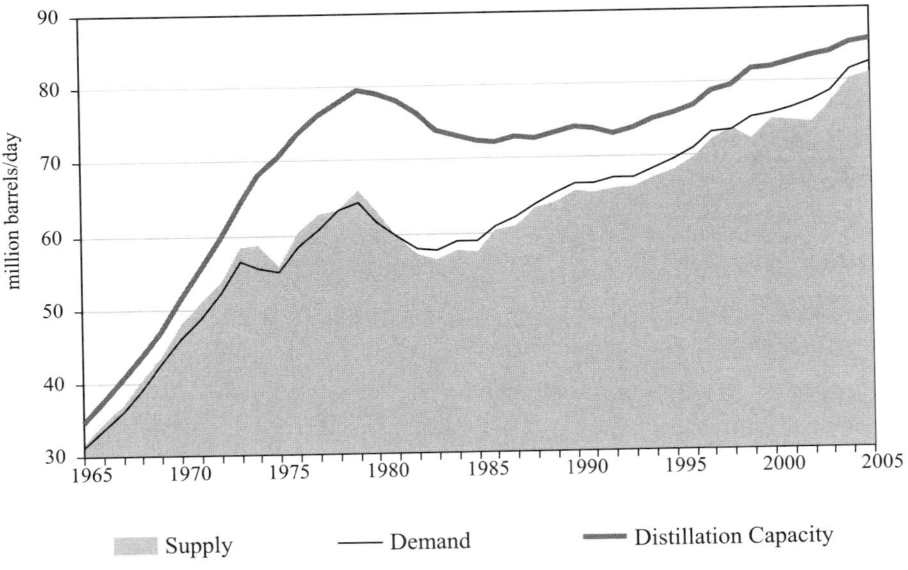

Source: IEA and EIA. Compiled by Repsol YPF Research Department

In a scenario of greater clean products demand, the shortage of conversion capacity has changed the economy of price differentials by disrupting the traditional evolution of the spread between light and heavy crude oil prices. Normally, the differential could be brought down by processing more heavy and medium oil in upgrading units. So what the wider spread indicates is a lack of conversion capacity (coking and hydrocracking) to process medium and heavy crude oils (see Figure 4.38).

Figure 4.38
Light/Heavy Crude Oil Differentials

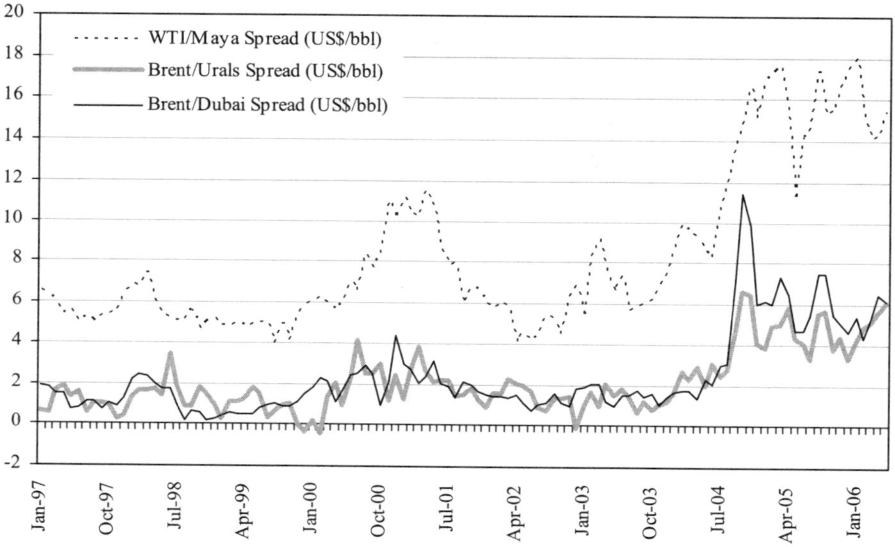

Source: Bloomberg and Reuters. Compiled by Repsol YPF Research Department

Over the years, upgrading units have been progressively modified to increase conversion rates and process heavier feedstocks. However, the upgrading capacity has stagnated in recent years after an increase in the early nineties. Figure 4.39 shows how conversion level (cracking and coking) has increased less than 0.5 million bpd since 1999, whereas, according to the IEA, demand for the transport sector (of clean products) has experienced an annual average increase of a 1.0 million bpd.

Figure 4.39

Variation in Total and Percentage of Conversion Capacity

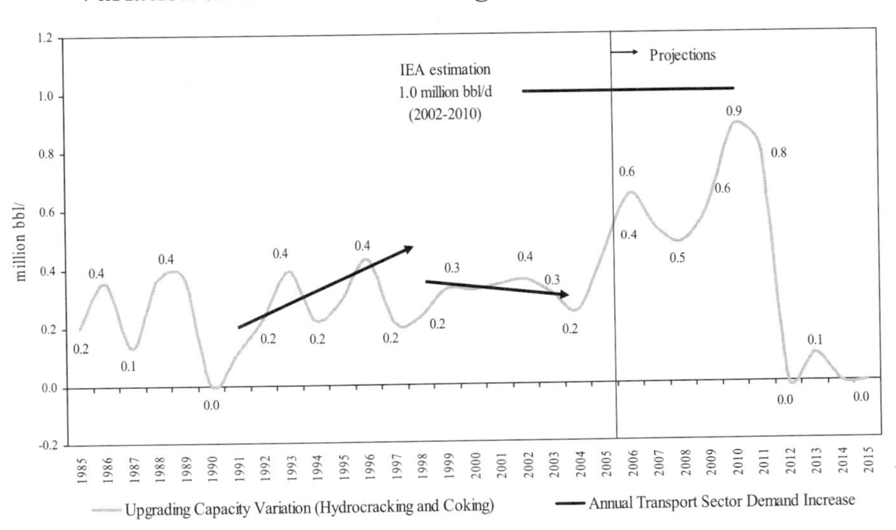

Source: *Oil & Gas Journal*, IEA and HETCO. Compiled by Repsol YPF Research Department

Expected Investments in Upgrading Refining and Regional Breakdown

The higher utilization of available refining capacity in the second half of the eighties (see Figure 4.40), and especially demand growth, prompted a larger investment in refining and an increase in refinery conversion capacity. The latter occurred mainly in Asia, but also in other OECD countries, making it possible to stabilize capacity in the 1992–2002 period. This situation began to change in 2003 and underwent a drastic alteration in 2004, when the capacity utilization rate rocketed.

In recent years, upgrading capacity has increased less than 0.5 million bpd by year (see Figure 4.40). Looking to the future, based on available data, it seems highly unlikely that the estimated increments will close the gap between the rising demand for clean products and the world's conversion capacity.

Figure 4.40
Geographical Breakdown of Refining Capacity Utilization

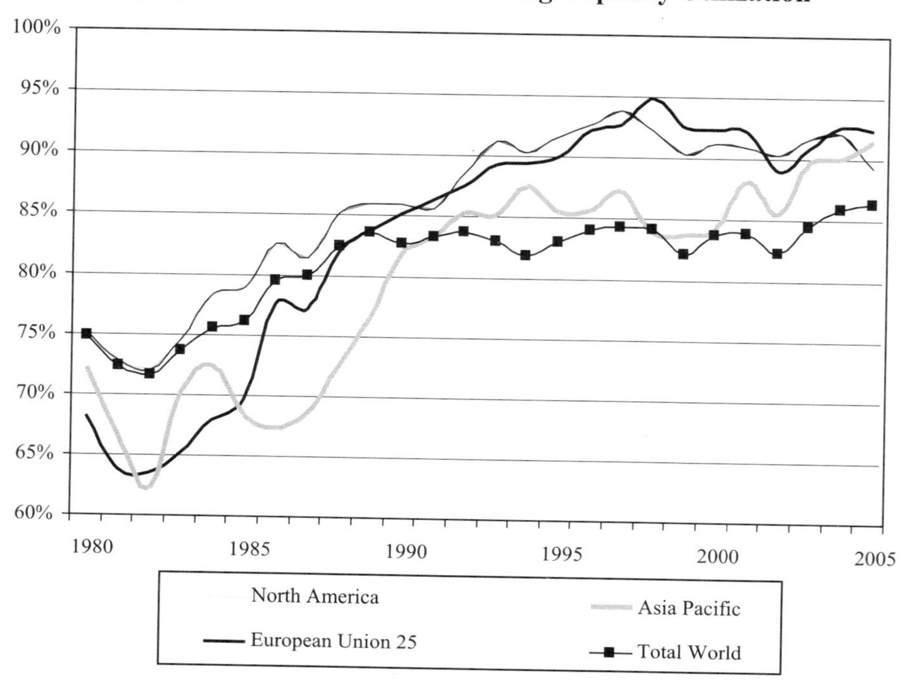

Source: *BP Statistical Review*. Compiled by the Repsol YPF Research Department

In short, when making a joint projection of the demand growth rates for clean products and the increases in refining capacity for these products, a supply shortage appears, which is set to last practically until 2012.

In this respect, excess demand and resulting high margins will be practically inevitable in view of the start up time necessary for new refineries or even new conversion units at existing facilities. According to data from various sources (see Figure 4.41), refining capacity will continue to grow at the current pace and supply will not meet demand until 2013.

Figure 4.41
Additional Refinery Upgrading Capacity Needed for Clean Product Production

Clean Product Production million bbl/d

28
27
26
25
24
23
22
21
20

2002 2003 2004 2005 2006 2007 2008 2009 2010 2011 2012 2013 2014 2015

Existing Capacity
Trendline Capacity Growth
Firm New Capacity Plans
Uncovered Requirement

Source: *Oil & Gas Journal* and HETCO. Compiled by Repsol YPF Research Department.

Regarding the expected upgrading capacity additions, Asia will be the region adding more capacity (see Figure 4.42). During the period 2006–2011, Asia will increase upgrading capacity by almost 1.2 million bpd, more than twice the capacity to be added by North America, the second region with greater additions (+ 0.5 million bpd).

Figure 4.42
Regional Upgrading Capacity Growth Forecast

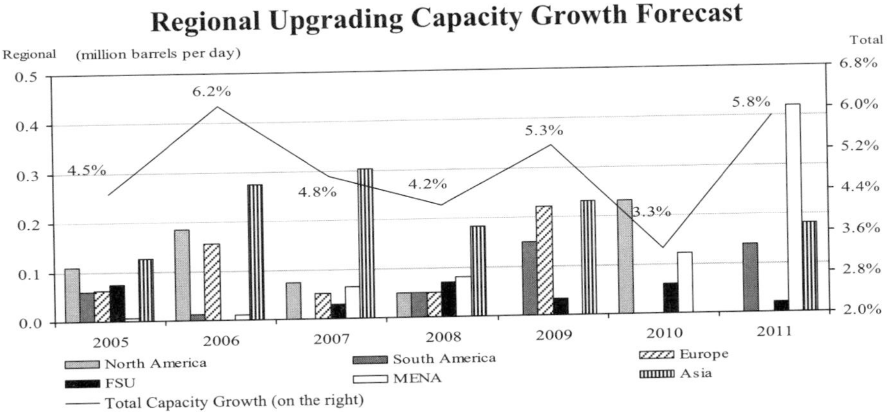

Source: *Oil & Gas Journal* and HETCO. Compiled by Repsol YPF Research Department.

However, as part of these expected capacity additions may be delayed or not even built, one has to be very prudent when making an assessment of the refining capacity additions. There are already analysts claiming that refining overcapacity will be built in the years to come. For the time being, however, the fact is that from 2005 to 2010 the total upgrading capacity to be added to the system would be around 3 million barrels per day. Assuming that global demand will increase at 1 million bpd annually, there would be about 6 million more barrels of demand till 2010. This would mean an upgrading deficit of about 3 million barrels per day by that year.

Another example could be the relationship between the expected increase in heavy crude oil production and the expected capacity additions of coking units. While over the period 2002–2009 the range of forecasts for incremental heavy crude goes from 1.5 million to more than 3 million barrels per day (as indicated earlier), the expected coking capacity additions are of 1.3 million barrels.

Figure 4.43
Regional Coking Capacity Growth Forecast

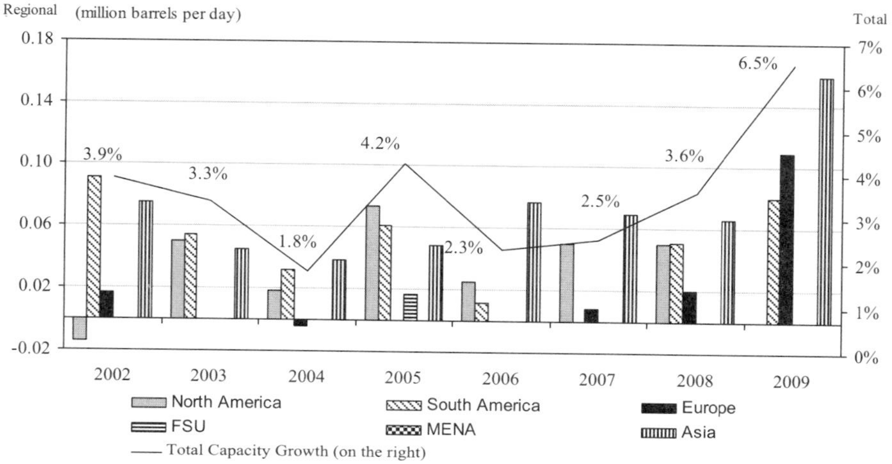

Source: *Oil & Gas Journal* and HETCO. Compiled by Repsol YPF Research Department.

Different Views on Investment in the Refining Sector

The need for greater investment in refining varies across the regions of the world. The regional imbalances in products are clear (see Figure 4.44). Europe, Asia and North America are net importers and the Middle East, North Africa, FSU and Sub-Saharan Africa are net exporters. In the last couple of years, only the Middle East has increased its exports significantly.

Figure 4.44
World Net Imports by Region (2005)

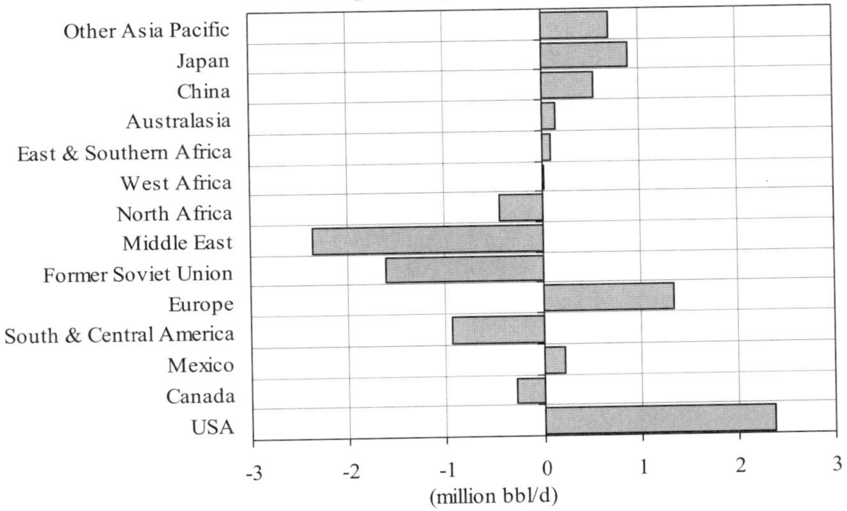

Other Asia Pacific
Japan
China
Australasia
East & Southern Africa
West Africa
North Africa
Middle East
Former Soviet Union
Europe
South & Central America
Mexico
Canada
USA
-3 -2 -1 0 1 2 3
(million bbl/d)

Source: *Oil & Gas Journal*. Compiled by Repsol YPF Research Department.

When considering the prospect of overinvestment or underinvestment in refining, it is worth enumerating the elements that provide incentives for investment and those which act as potential brakes.

On the positive side, it may be noted that there is a profitable opportunity for more investment in upgrading if OPEC production increases its market share. Also, the upward evolution in refining margins will surpass the expected cost increase in refinery building, and therefore adding upgrading units to current refineries will prove to be good business.

On the contrary, one can say that a substantial part of the investment has to go towards meeting the new and more stringent requirements in terms of sulfur content. That would imply high differentials between sweet (which are normally light) and sour (which are normally heavy) crude oils and therefore high conversion margins.

Regarding the brakes on investment, it ought to be mentioned that overcapacity was built in the past and the historic profitability of the sector is low. Moreover, firms are penalized in their share value if they enter into an investment that is not highly profitable according to calculations based on investment bank forecasts for refining margins. It may also be argued that even after taking into account the average high margins of the last couple of years, building grassroots refineries is not a profitable investment.

Last but not the least, independent of any profitability considerations, a refinery can be built by national companies following government objectives such as "assuring security of product supply" or "increasing the value added of oil activities." If the chances of such "national interest" investment rise, the potential risk of overinvestment also increases.

1-Positive Incentives for Investment

The only way to evaluate the profitability of investing in refining is by analyzing whether refining margins are going to be high in the future. This section analyzes the factors that encourage investment in refining and consequently favor high refining margins.

PROSPECTIVE CONTINUATION OF TRENDS: OPEC PRODUCTION INCREASES

The continuation of the current trend, based on increasing medium/ heavy crude oil and lighter demand forms the basis for the most feasible future scenario regarding global oil. The main facts supporting this hypothesis, on the supply side, are the increasing OPEC production needs and the current OPEC petroleum quality, both of which were discussed earlier.

Regarding the increasing OPEC production needs, in the "Medium-Term Oil Market Report" issued in July 2006, the IEA states that the "Call on OPEC crude" will increase 1.4 million bpd in the period from 2006–2011. Hence, even with the major increase of non-OPEC supply (+5.4 million bpd), OPEC production has to rise in order to meet global demand and it could be expected that the quality of its future production be medium/heavy and sour.

The IEA projections point to an increase in OPEC production but could this growth have any influence over refining margins and particularly over deep conversion refining margins? Analyzing the relation between refining margins, crude oil differentials and OPEC production could help to answer this question.

As discussed earlier, the differential between light and heavy crude oil prices is an indicator of the behavior of refining margin performance. Therefore, if OPEC production influences directly light/heavy crude oil price differential, it can be assumed that OPEC production and deep conversion margins are related.

The Maya Coking margin and the related price differential between West Texas Intermediate (WTI) and Maya crude oils are considered for the analysis. As seen in Figure 4.44, these two variables are highly correlated.

On the other hand the WTI/Maya crude oil price differential is related to OPEC production too. As seen in Figure 4.45, during the period 2001–2006, both variables have evolved closely, which could be a first indicator of the relationship between OPEC production and light/heavy crude oil price differential. However, the high correlation says nothing about causality relationship between these two variables.

The existence of a causality relationship between OPEC production and WTI/Maya spread would help to determine that OPEC production is a driver for deep conversion margins and that supply conditions regarding quality are relevant. In addition, as discussed earlier and shown in Appendix-I, the WTI/Maya spread explains the coking margins.

Figure 4.45
Relationship between OPEC Production and Crude Oil Price Differential

OPEC Production (million bbl/d) on the left
WTI/Maya Differential (US$/bbl) on the right
Correlation coeficient = 0.88

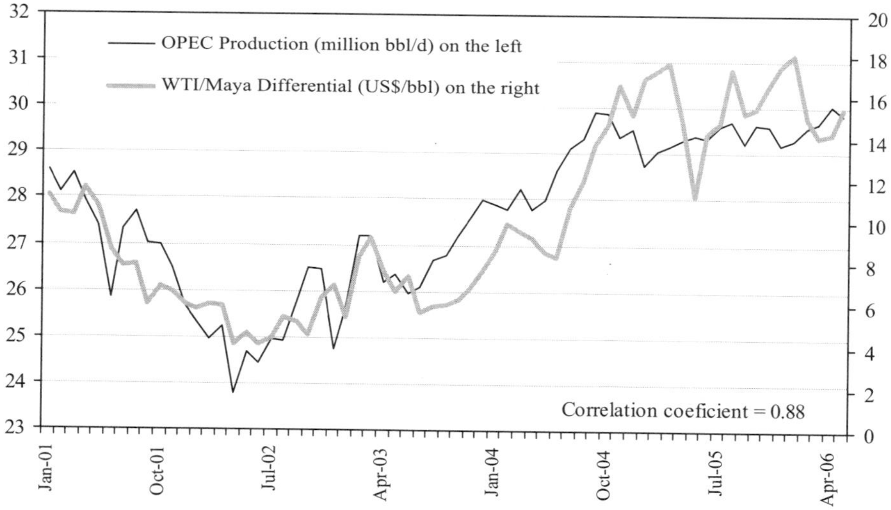

Source: IEA, Bloomberg and Reuters. Compiled by Repsol YPF Research Department.

In order to determine whether or not OPEC production causes or affects the light/heavy crude oil price differential, a Granger Causality Test was carried out (see Appendix-II for the complete analysis).

The outstanding conclusion is that OPEC production does cause the crude oil price differential and not the opposite. Therefore, according to the results in Appendix-II, the expected evolution of OPEC production in the short and medium term is an indirect or implicit indicator of coking margin levels, and consequently, if forecasts of OPEC production are right, then investing in deep conversion will be very favorable.

Nevertheless, the timing for investing in deep conversion is very relevant. If investment takes place too fast, the crude differential and the margin will shrink and the competitive advantage for complex refineries will disappear. Being aware of this aspect is crucial for carrying out profitable projects.

NEEDED INVESTMENT IN HYDRO-TREATING AND MARGINS LEVELS

The hydrotreating process cannot be omitted in the analysis of refining sector bottlenecks. The trend towards more stringent environmental regulation and particularly on sulfur content limit (see Figures 4.46 and 4.47), will have many economic effects regarding refining margins. First, it will divert investment from upgrading capacity projects, and this will diminish the risk of overinvestment in conversion. Second, it should keep margins for desulfurization relatively high as new requirements come into force. Finally, in operational terms, the increase in hydrotreating will mean more energy consumption and a reduction of product yield of more than 3.5 percent, which is equivalent to a reduction of refinery capacity.

Figure 4.46
Sulfur Content Specifications for Gasoline as per Implementation Schedule

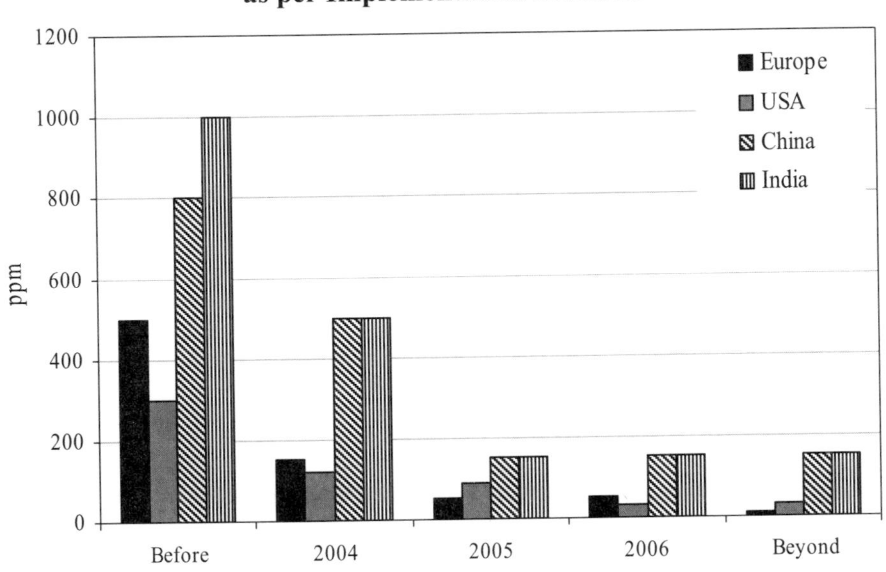

Source: *Oil & Gas Journal*. Compiled by Repsol YPF Research Department.

Figure 4.47
Sulfur Content Specifications for Diesel as per Implementation Schedule

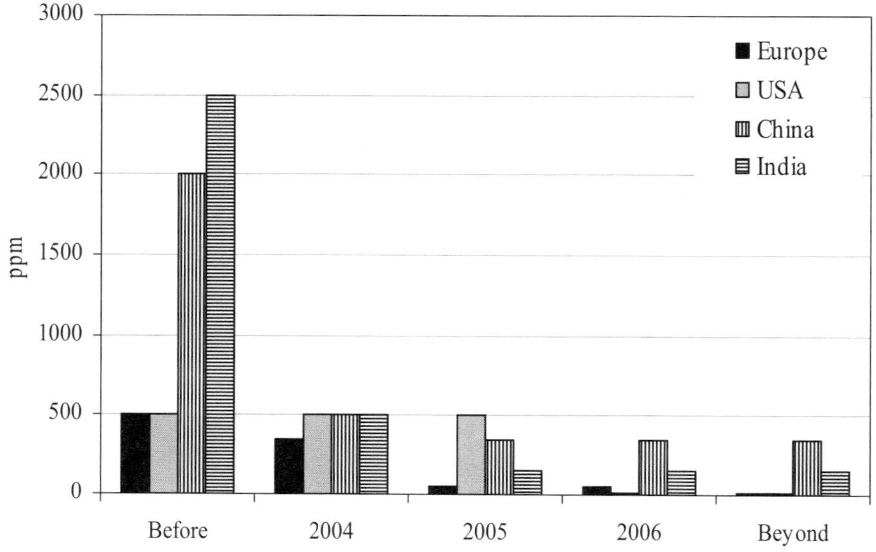

Source: *Oil & Gas Journal*. Compiled by Repsol YPF Research Department.

All the previous arguments will contribute to a tight refining situation, particularly when the average sulfur content of future crude oil is expected to increase as average crude oil quality gets sourer (as seen earlier in Figure 4.9).

The new requirements, particularly in China and India, but also in most of the emerging economies, are very demanding in terms of capital.

The magnitude of the investment effort in the emerging world is at least challenging. On the one hand, oil consumption is growing fast and hydrotreating capacity, as a percentage of crude capacity, is very low outside the OECD area, especially in China and India (see Figure 4.48). These two countries together represent less than 5 percent of hydrotreating global capacity. The situation could be even more challenging if oil production continues to become sourer.

Figure 4.48

Hydrotreating as Percentage of Domestic Crude Capacity

100%
80%
60%
40%
20%
0%

98% Japan
78% USA
66% Europe
58% World
49% Others
45% India
41% Middle East
19% China

Source: *Oil & Gas Journal* and HETCO. Compiled by Repsol YPF Research Department

The new requirements coming into force in the sector have prompted hydrotreating capacity expansions. Over the period 2005–2011, Asia leads capacity additions followed by North America and South America (see Figure 4.49).

Figure 4.49

Regional Hydrotreating Capacity Growth Forecast

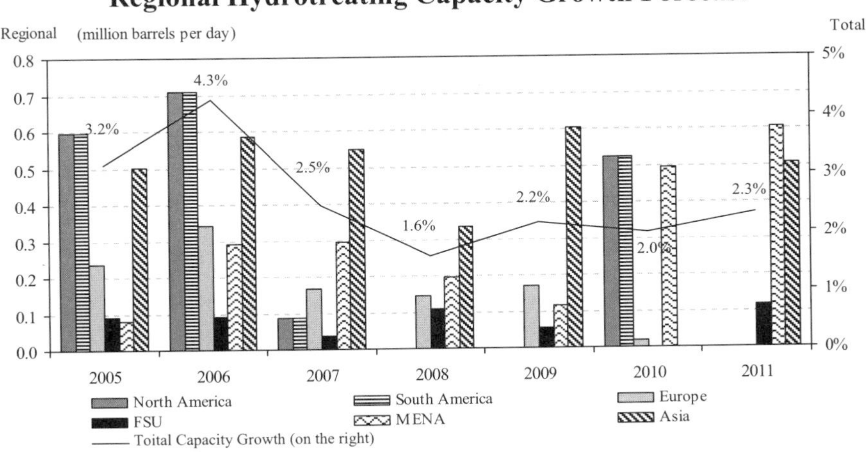

Source: *Oil & Gas Journal* and HETCO. Compiled by Repsol YPF Research Department

EXPECTED COST OF METALS VIS-À-VIS OIL

Another important factor to consider in evaluating whether there will be a significant increase in refining conversion capacity, is the role played by metal prices as one of the main determinant "costs" of refinery investment.

The price level of metals in the coming years will be a fundamental variable to assess the costs and profitability of any investment in refining. A scenario of high metal prices implies that investment in conversion would not reach the profitability threshold until refining margins had amply surpassed historical levels.

A key question is whether the higher costs involved in building refineries or new conversion units, stemming from the current and future price of metals used for capacity increases, could discourage investment or make it scarcely profitable.

Although it is well known that a large number of factors determining metal and oil prices coincide, certain elements would seem to indicate that metal prices will follow a more moderate trend than refining margins over the next few years.

In fact, a considerably slower trend is emerging in the prices of certain metals (such as lead, nickel and especially brass) that are beginning to fall. In contrast, the price of metals with important restrictions on the supply side (copper and iron) continues to rise sharply.

The most significant aspect of this trend is that oil prices will balance out at a higher level than metal prices (see Figure 4.50). Therefore, it is expected that metal price index levels will fall below the oil price and this will be positive for the refining industry in terms of lower cost pressures.

Figure 4.50
Metal and Oil Price Index (1995 = 100)

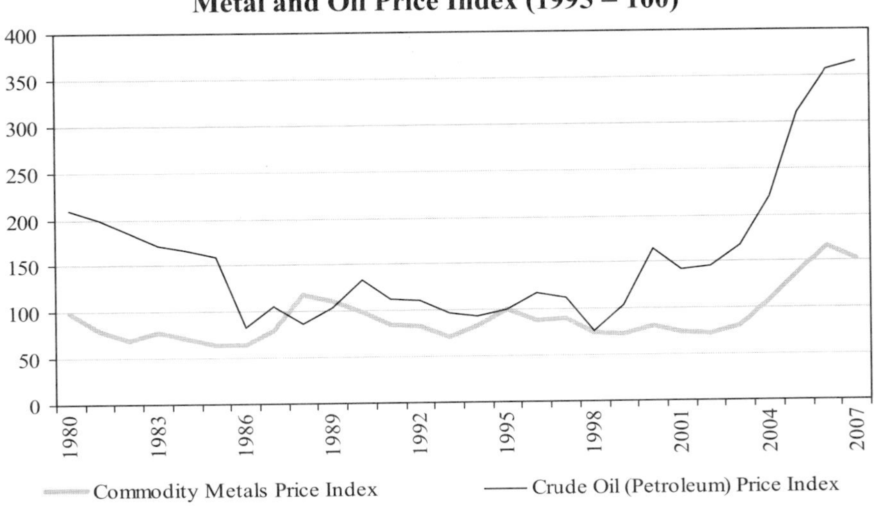

Note: The Commodity Metals Price Index includes copper, aluminium, iron ore, tin, nickel, zinc, lead, and uranium price indices. The Crude Oil (Petroleum) Price Index is a simple average of three spot prices: Dated Brent, West Texas Intermediate and the Dubai Fateh.

Source: International Monetary Fund (IMF). Compiled by Repsol YPF Research Department.

2- Factors Discouraging Investment in Refining

REFINING MARGINS AND INVESTMENT DECISIONS

The main argument as to why overinvestment in the refining sector will not occur is that the industry is capital intensive, requires long term investments and refining margins have been historically low and volatile.

To assess the profitability of such investments, it is necessary to make "future refining margin assumptions." In the refining industry these "assumptions" continue to be highly determined by past events and historic margins. The adaptation to a new scenario of high margins is not easy in terms of investments, if those responsible for the investment decisions are familiar with the earning volatility of the industry and the historical overcapacity.

New refineries are only reasonable with high refining margin projections, in order to gain a profitable return on capital. Even now, the case for some investment is not that clear. Even when using the last two years average refining margin as "future refining margin assumptions" the expected profitability of a new grassroots refinery is not absolutely evident to some oil analysts.

While there are different views on the potential profitability of a new refinery, there is wide consensus (using the same "future refining margin assumptions" mentioned in the previous paragraph) that the building of upgrading units at existing refineries is a profitable business. In this case, the caveat should be whether or not there are other more profitable alternative investments.

UNDERESTIMATION OF CURRENT MARGINS BY FORECASTERS

Another factor that explains why recent investment in refining has been lagging behind demand growth is the systematic difference between crude oil prices and refining margin projections or forecasts and current data. Looking at the crude oil price projections for 2003, 2004 and 2005 carried out by investment banks in January each year, the average difference with regard to the actual annual oil price, has been about 10 dollars per barrel.

Figure 4.51

Refining Margins: Investment Bank Projections and Real Data (2004)

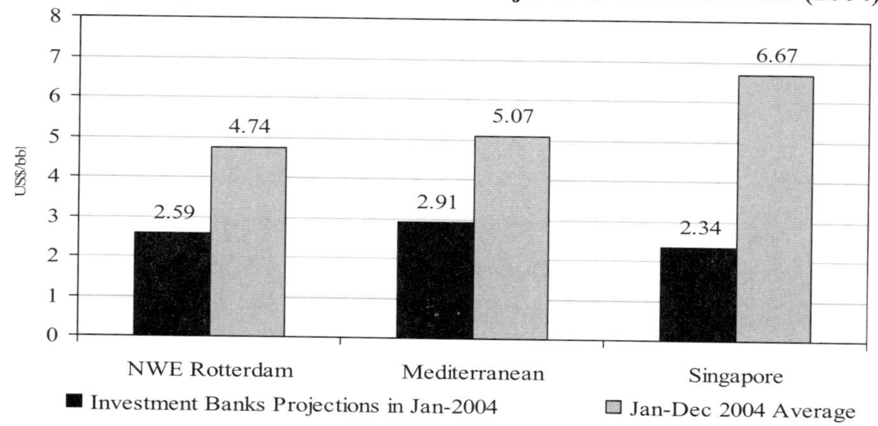

US$/bbl

8
7
6
5
4
3
2
1
0

2.59
4.74
2.91
5.07
2.34
6.67

NWE Rotterdam
Mediterranean
Singapore

■ Investment Banks Projections in Jan-2004
□ Jan-Dec 2004 Average

Source: Merrill Lynch, Deutsche Bank, JP Morgan, Morgan Stanley, Lehman Brothers and Reuters. Compiled by Repsol YPF Research Department

This underestimation of the crude oil price is mirrored in the refining margins. Thus, the refining margin projections carried out by the same investment banks in January of 2003, 2004 and 2005 were significantly different from the real data at the end of these years. This difference is illustrated by the 2004 price projections and real data as shown in Figure 4.51.

Even those investors more bullish on the profitability prospects in the refining sector in recent years, have faced constraints in raising money for their projects. No investment bank foresaw high refining margins and thus they were not in a position to lend money based on such forecasts.

Concluding Remarks

The current challenges facing refining may be summarized in one sentence: "The last barrel produced will be medium/heavy and sour, while the last barrel demanded will be light (clean)."

Effectively, the demand side is characterized by an increasing need for clean products. Demand is being driven by strong economic growth, the growing participation of the transport sector in total final consumption and by the challenge posed by new environmental requirements. Behind demand resilience is the fact that, both now and in the medium term, there is no effective substitute for gasoline and diesel to fuel the global vehicular motion. With regard to the new environmental restrictions, although not all regions start from the same experience level, all have passed legislation to further reduce sulfur content in final oil products.

On the supply side, no analyst expects an improvement in oil quality. No more than 60 percent of total oil production will continue to be medium and heavy for the foreseeable future. Moreover, if long term IEA forecasts of oil quality are considered or implicit projections are done on the basis of long term forecasts of OPEC market share, then the expected oil quality will further deteriorate.

What is more consensual is the fact that sulfur content will increase. Even the most optimistic observers predict a significant rise of sulfur content in the average global supply. By and large, it is an undeniable fact that supply is coming from regions characterized by their medium/heavy and sour petroleum (OPEC) and if these very petroleum reserves are mostly medium and sour, what will the expectations about future production be? The answer seems to be clear. Crude oil quality is evidencing a deterioration which, even if it is considered "slight," could nevertheless denote a significant challenge for a refining industry without adequate conversion (upgrading) capacity to meet the clean demand requirements.

The shortage of additional upgrading capacity has been a real bottleneck for refining over the last years. In fact, the conversion level (cracking and coking) has increased less than 0.5 million bpd since 1999 whereas demand for the transport sector has experienced an annual average increase of a 1.0 million bpd. In addition, although over the period 2002–2009 the range of forecasts for incremental heavy crude goes from 1.5 million to more than 3 million bpd, the expected coking capacity additions are of 1.3 million barrels.

Making a joint projection of the demand growth rates for clean products and the increases in refining capacity for such products, it appears that supply will not meet demand until 2013. This imbalance can only be solved by investing in refining (conversion units), but investment should be based on a high margin outlook.

In this respect, some factors discouraging investment are that refining margins have been historically low and volatile, and there is a systematic difference between crude oil price and refining margin forecast and current data.

It has been econometrically proved in Appendix-I of this chapter that there is a causality relation (according to a Granger Causality Test) from crude oil price differentials to refining margins. In particular, it has been proved that the light/heavy crude oil price differential (WTI–Maya) is an explanatory variable of the high conversion margin (Maya Coking).

Therefore, the wider the price differential between light and heavy crude oils, the higher is the conversion margin (See Appendix-I).

It has also been tested through a Granger Test that OPEC production does cause the light/heavy crude oil price differential: WTI–Maya (See Appendix-II). In this respect, considering that the need for OPEC crude is expected to increase and its petroleum quality is medium/heavy and sour, crude oil price differentials will remain high and investing in conversion units should remain profitable. Nevertheless, the timing for investment in deep conversion units can be crucial.

Moreover, a substantial part of investment must go towards meeting the more stringent requirements in terms of sulfur content. This investment will divert funding from conversion units, diminish the risk of overinvestment in conversion and will contribute to a tight refining situation by reducing refinery yields.

APPENDIX-I

It is econometrically indicated in this Appendix that there is a causality relation (according to a Granger Causality Test) from crude oil price differentials to refining margins. In particular, it is indicated that the light/heavy crude oil price differential (WTI–Maya) is an explanatory variable of the high conversion margin (Maya Coking).

Table 4.6
Variable Glossary

Variable	Description	Transformation*
SPREAD	WTI–Maya price spread	Differential (D), Logarithmic (L)
COKING	Maya Coking refining margin	Differential (D), Logarithmic (L)

* Transformations which are necessary to obtain stationarity

Source: Repsol YPF Research Department.

The causality relationship between the WTI–Maya spread and the Maya Coking refining margin is investigated by applying econometric methods to monthly data for the period January 1997 to May 2006. Both data series are integrated of order one I (1) (unit root test). By using the Johansen Cointegration Test, it is concluded that there is a cointegrating relation between the WTI–Maya spread and the Maya Coking refining margin.

The Granger causality test may be applied as follows:

Table 4.7
Granger Causality Test

Null Hypothesis:	Obs	F-Statistic	Probability
DLSPREAD does not Granger Cause DLCOKING	108	3.14864	0.01755
DLCOKING does not Granger Cause DLSPREAD		1.21959	0.30739

Source: Repsol YPF Research Department.

It can be concluded that:

- The WTI–Maya spread Granger-causes the Maya Coking refining margin.

- The Maya Coking refining margin does not Granger cause the WTI–Maya spread.

Based on this information, the cross correlations, the univariate analyses and Akaike and Schwarz´s information criteria, a VEC (1) model is estimated with the following results:

Table 4.8
VEC (1) Model

Cointegrating Eq:	CointEq1	
LSPREAD (-1)	1.000000	
LCOKING (-1)	-0.598444 (0.03943) [-15.1759]	
C	-1.125837	
Error Correction:	D(LSPREAD)	D(LCOKING)
CointEq1	0.002115 (0.07093) [0.02982]	1.017253 (0.17354) [5.86185]
D(LSPREAD(-1))	-0.153949 (0.11244) [-1.36916]	-0.468372 (0.27510) [-1.70257]
D(LCOKING(-1))	-0.032526 (0.04124) [-0.78880]	0.105200 (0.10089) [1.04276]
C	0.009708 (0.01282) [0.75738]	0.014193 (0.03136) [0.45258]
R-squared	0.041616	0.266074
Adj. R-squared	0.014745	0.245496
Sum sq. resids	1.944429	11.63901
S.E. equation	0.134804	0.329812
F-statistic	1.548749	12.93039
Log likelihood	66.97101	-32.34034
Akaike AIC	-1.134613	0.654781
Schwarz SC	-1.036972	0.752422
Mean dependent	0.008071	0.012682
S.D. dependent	0.135809	0.379695
Determinant resid covariance (dof adj.)		0.001663
Determinant resid covariance		0.001545
Log likelihood		44.23764
Akaike information criterion		-0.616894
Schwarz criterion		-0.372793

Source: Repsol YPF Research Department.

From the VEC model results it can be concluded that the WTI–Maya spread is an explanatory variable of the Maya Coking refining margin. However, the inverse relation does not exist.

With a cointegration relation, the Granger causality test loses reliability. So, it is necessary to do a causality relation test, through the Wald test on the VEC model that will allow us to test the causality relation between variables in the long and short term.

Table 4.9
Wald Test

Wald Test:			
Equation: Untitled			
Test Statistic	Value	df	Probability
F-statistic	17.65770	(2,107)	0.0000
Chi-square	35.31540	2	0.0000
Null Hypothesis Summary:			
Normalized Restriction	(=0)	Value	Std. Err.
C(1)		1.017253	0.173538
C(2)		-0.468372	0.275097
Restrictions are linear in coefficients.			

Source: Repsol YPF Research Department.

The null hypothesis is rejected, indicating that both long and short term causality exists.

The impulse response is made with one standard deviation.

There is a significant response of the Maya Coking refining margin to a WTI–Maya spread innovation, while there is no significant response of the WTI–Maya spread to a Maya Coking refining margin innovation.

Figure 4.52
Response of LSPREAD to LCOKING

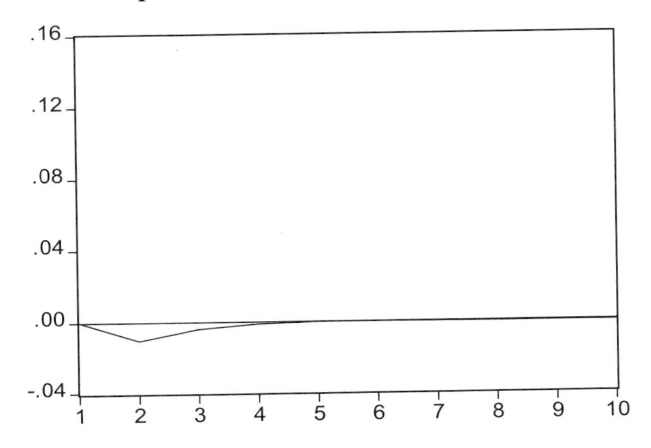

Source: Repsol YPF Research Department.

Figure 4.53
Response of LCOKING to LSPREAD

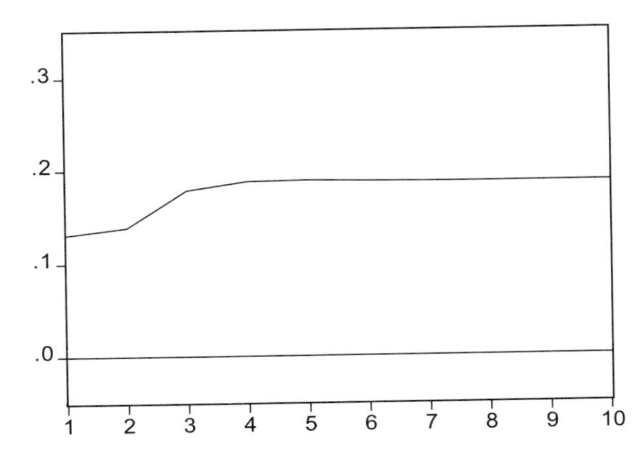

Source: Repsol YPF Research Department.

APPENDIX-II

In this Appendix it is shown through a Granger Test that OPEC production does cause the light/heavy crude oil price differential: WTI-Maya.

Table 4.10
Variable Glossary

Variable	Description	Transformation*
SPREAD	WTI–Maya price spread	Differential (D), Logarithmic (L)
OPEC	OPEC production	Differential (D), Logarithmic (L)

* Transformations which are necessary to obtain stationarity
Source: Repsol YPF Research Department.

The causality relationship between the WTI–Maya spread and the OPEC production is investigated applying econometric methods to monthly data for the period January 1997 to May 2006. Both series data are integrated of order one I (1) (unit root test). By using the Johansen cointegration test, it is concluded that cointegration is absent.

In absence of cointegration, the Granger causality test may be applied as follows:

Table 4.11
Pairwise Granger Causality Tests

Null Hypothesis	Obs	F-Statistic	Probability
DLSPREAD does not Granger Cause DLOPEC	110	1.14	0.32
DLOPEC does not Granger Cause DLSPREAD		3.48	0.03

Source: Repsol YPF Research Department.

It may be concluded that:

- The WTI–Maya spread does not Granger cause OPEC production.
- OPEC production Granger causes the WTI-Maya spread.

Based on this information, the cross-correlations, the univariate analyses and Akaike and Schwarz's information criteria, a VAR (1) model is estimated with the following results:

Table 4.12
VAR (1) Model

	DLOPEC	DLSPREAD
DLOPEC(-1)	-0.218441	1.338532
	(0.09369)	(0.50282)
	[-2.33158]	[2.66205]
DLSPREAD(-1)	-0.027508	-0.228239
	(0.01736)	(0.09316)
	[-1.58467]	[-2.44990]
C	0.001171	0.008441
	(0.00231)	(0.01241)
	[0.50655]	[0.68026]
R-squared	0.081182	0.093496
Adj. R-squared	0.064167	0.076709
Sum sq. resids	0.063850	1.839172
S.E. equation	0.024315	0.130497
F-statistic	4.771142	5.569493
Log likelihood	256.5691	70.05975
Akaike AIC	-4.568813	-1.208284
Schwarz SC	-4.495582	-1.135053
Mean dependent	0.000775	0.008071
S.D. dependent	0.025135	0.135809
Determinant resid covariance (dof adj.)		9.62E-06
Determinant resid covariance		9.10E-06
Log likelihood		329.1685
Akaike information criterion		-5.822856
Schwarz criterion		-5.676395

Source: Repsol YPF Research Department.

From the VAR model results it may be concluded that the OPEC production is an explanatory variable of the WTI-Maya spread. However, the inverse relation does not exist.

The impulse response is made with one standard deviation:

Figure 4.54
Response of DLOPEC to DLSPREAD

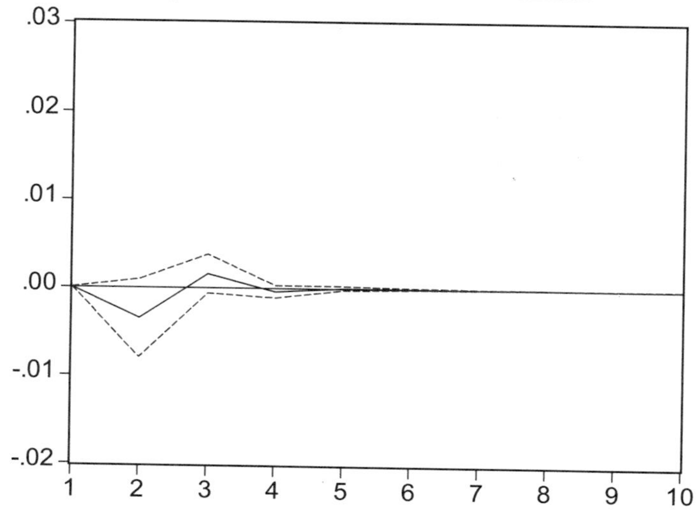

Source: Repsol YPF Research Department.

Figure 4.55
Response of DLSPREAD to DLOPEC

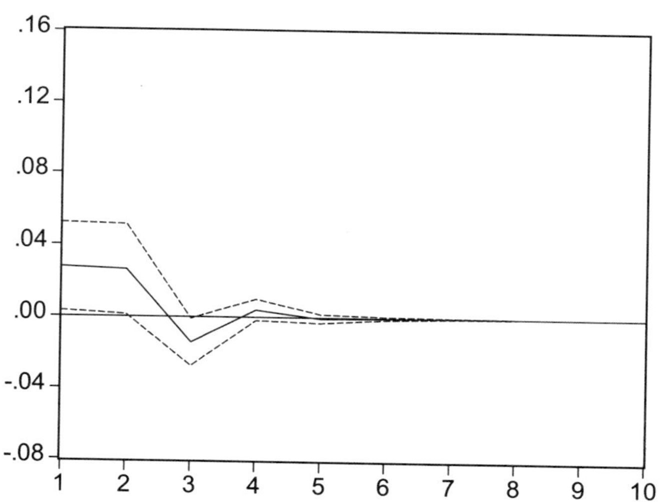

Source: Repsol YPF Research Department.

There is a significant response of the WTI–Maya spread to an OPEC production innovation, while there is no significant response of OPEC production to a WTI–Maya spread innovation.

5

Investment Prospects for Foreign Oil Companies: Risks and Potential

Jean-Pierre Favennec

The high price of crude oil is primarily the result of saturated production capacities.[1] Substantial capacity surpluses had built up in the 1980s, particularly in the OPEC countries, in the aftermath of the oil crises of 1973 and 1979, as a result of falling demand and expanding production in non-OPEC areas. More recently, with rising oil consumption, these capacity surpluses were gradually reduced and due to the sharp increase in demand especially in 2004, have fallen currently to very low levels. Considerable investment is needed, not only in this sector but also in the transport and refining sectors. What are the levels of investment required? In which countries should these investments be made? Who will make the investments? A detailed analysis is needed into all these aspects, but the role of the Gulf countries, where two-thirds of the world's oil reserves are located, will undoubtedly be decisive.

Hydrocarbons: The Current World Situation

At the close of the 20th century, energy seemed abundant and generally available at cheap rates. As a result of actual or potential capacity surpluses, consumers enjoyed relatively low prices, with oil prices ranging from US$15–25 per barrel and falling at times to very low levels, and gas prices, which are indexed to oil prices, remaining low as well.

However, since 2000, the energy picture has altered radically as oil prices hit record highs and gas prices in the United States reached unprecedented levels. What are the main factors at the root of the current situation? Is it due to an unexpected rise in demand, inadequate production capacity or deregulation of the energy sector?

Energy Consumption and Economic Development

Energy consumption has grown sharply over both the long and short terms, despite the fact that the energy consumption in general, and oil consumption in particular, were directly impacted by the major economic and political events of the latter part of the 20th century: the oil crises of 1973 and 1979, the collapse of oil prices in 1986, the dissolution of the Soviet Union and the severe economic crisis that struck one of the world's leading energy producers and consumers, as well as the 1997–98 monetary crisis in emerging countries, particularly in Asia.

As a percentage of all energy needs, oil covers 35 percent,[2] while coal accounts for 23 percent, natural gas constitutes 21 percent, nuclear energy amounts to 4 percent and hydroelectric power comprises 4 percent. Biomass energy (which amounts to 12 percent of energy production) remains a major source of fuel in many African countries. Despite the appeal of new, renewable energy sources (such as solar, wind and geothermal), these account for less than 1 percent of all energy consumed.

In the past two years, oil consumption has risen much more rapidly, driven by worldwide economic growth at record levels. From less than 2 percent per year, the growth rate reached 3.5 percent in 2004 with this rising trend continuing in 2005. Gas consumption has increased at a similar rate.

Oil and Gas Production

Oil production increased tenfold between World War II and the present day. Although historically concentrated in North America, production spread to regions such as the Middle East and East Asia and later to North Africa, sub-Saharan Africa and the North Sea region.

The Middle East is the world's leading oil producing region. Since the early 1990s, Saudi Arabia has been the top-ranking oil producer worldwide. However, its production of approximately 10.5 million barrels per day is linked to OPEC decisions pertaining to an increase or decrease in the Kingdom's production quota as a member state.

OPEC's goal of maintaining a balance between supply and demand to ensure sufficiently high prices for all its member countries will be easier to achieve in view of the group's high overall market share. This market share peaked at more than 50 percent in the early 1970s, fell to less than 30 percent in the mid-1980s, and now stands at approximately 40 percent.

Until recently, natural gas production varied according to demand. The appearance of "gas bubbles," especially in the United States in the 1980s, led at times to very low prices. Gas resources are substantial and should be more than adequate to meet demand. However, the remote location of gas fields to be brought into production (northern Russia and the Middle East) will require considerable investment, for which producers are now requesting guarantees (such as "take or pay" contracts) that consumers are not always willing to provide.

Prices and Production Capacity

1-Oil prices

After dropping to US$10 per barrel in late 1998, crude oil prices have increased sharply, hitting a new record in 2006 at more than US$70 per barrel (It may be noted, however, that prices in 1980, following the second oil crisis, were close to US$90 in 2006 dollars). Some of the explanations given for this situation are listed below:

- diminishing surplus production capacity
- strong demand for oil due to sharp economic growth (particularly in China and the United States)

- the effect of "speculation" (although none deny its reality, opinions vary widely as to its impact)
- saturation of existing refining capacity (the crude oil in the market is increasingly heavy, while light oils are needed to produce maximum amounts of gasoline and diesel oil.

As a result, oil prices remain high and by all accounts, are higher than the minimum price needed to ensure that the operational chain runs smoothly and the oil market is adequately supplied. As a matter of fact:

- Per barrel production cost is no more than US$3 in Saudi Arabia, US$5 in the Middle East as a whole, US$5–10 in the other producer countries and US$15 for the barrels that are most costly to produce: oil from the difficult regions of the North Sea and synthetic crude oil from the Orinoco and Athabasca regions.
- Most OPEC member countries (with the exception of Indonesia) depend on oil for 80–90 percent of their budgetary revenues.[3] They traditionally draw up their budgets with a barrel price of US$20 to US$30.
- The decisions of oil companies on whether to bring a field into production are based on a profitability analysis which uses a price of US$20–25[4] per barrel of crude oil.

The high price should serve to limit the rise in demand and encourage producers to bring new fields on stream. However, until 2006, such consequences had failed to materialize fully. Why?

On the demand side, whereas the two oil crises – which caused prices to increase tenfold – triggered a 15 percent decrease in demand, the high prices observed in the past five years or so (with the exception of the post-9/11 period) seem to have had little effect on reducing demand. Several explanations have been offered:

- Oil has far less importance in today's economy than it did 20 years ago, particularly because of more efficient use and an increased share of the services sector (which uses little energy) in domestic output.

- The high proportion of taxes included in gasoline and diesel fuel prices (in most consuming countries) considerably reduces the impact of oil price fluctuations.

On the supply side, non-OPEC production seems to be leveling off in many countries, with the exception of the Commonwealth of Independent States (CIS). These countries include Russia and Caspian Sea states (particularly Kazakhstan and Azerbaijan) and West Africa. Only the OPEC countries, and especially those of the Middle East, have high potential to increase their production in the future. Saudi Arabia has consistently expressed its desire to have surplus production capacity of 1.5–2 million barrels per day and claims to be willing to gradually increase its total production capacity to 12 or even 15 million barrels per day if the market so demands.

2-Natural gas prices

The transport cost of crude oil is low compared to its final price. Thus, for oil shipped from the Middle East at a price of US$60 per barrel, transport costs to the United States or Japan are no more than US$2 or US$3 per barrel. However, the transport cost of gas represents a very high percentage of its final price. As a result, there is not one world gas market, but three regional markets: North America, Europe and Northeast Asia.

For many years, gas prices in North America remained low because production was high and adequate to satisfy local needs. On the other hand, Europe became dependent very quickly on imports from Russia, Algeria and more recently, from Norway. Northern Asia and particularly Japan, the world's largest importer, imports all its gas from Indonesia, Malaysia, Australia and the Middle East. Historically, gas prices have been higher in Europe than in North America and higher in Asia than in Europe, which led to costly imports of LNG.

In the first years of the century, the situation changed with the sharp rise in gas prices in the United States. Fast-growing demand could no longer be satisfied by production which, despite exploration for new fields, was tapering off. Thus the need arose to reduce demand and allow imports of LNG. Regasification terminals built in the United States in the 1980s and little used until now are being utilized, and many more projects are on the agenda.

Hydrocarbon Consumption Forecast[5]

Worldwide energy consumption will continue to expand in the coming decades as a result of economic growth (3.2 percent per year) and population growth (1 percent per year).

Several future scenarios are possible, depending on the rate of economic growth and the desired degree of environmental protection. For example, worldwide consumption in 2020 could amount to 12–15 Gtoe (versus 10.2 Gtoe in 2004). Although the situation in 2050 is very difficult to predict, energy requirements may have doubled by that date. Most scenarios, at least until the year 2020, share a number of common points:

- Energy intensity decreases by 1.5 percent per year
- Annual growth in global energy demand is 1.7 percent
- Fossil energy sources continue to account for over 80 percent of commercial energy supply
- Proportion of hydrocarbons in commercial energy consumption remains at 60 percent
- Oil continues to account for approximately 35 percent of this commercial consumption.

Oil will remain the predominant source of energy. As a percentage of total energy sources, it will change very little, and oil consumption could exceed 100 million barrels per day around 2020, compared with 84 million barrels today. It will be used primarily for transport.

The worldwide demand for gas could almost double between now and 2030, reaching 4.9 trillion cubic meters a year by that time. The market share of gas, a clean fuel, will see the highest rate of growth among energy sources, relegating coal to third place.

Increased Dependence of High-Consumption Regions

The gap between production areas and consumption areas will become more pronounced. The increased demand for energy will be concentrated mainly in regions that are already high-consumption zones where hydrocarbon reserves are being rapidly depleted. Increased trade will therefore be needed to accommodate the rise in demand. During this period, the developing countries, headed by India and China, will begin to consume more energy than the developed countries.

United States

The United States is by far the leading consumer and largest importer of energy in the world. With only 5 percent of the world's population, this country uses nearly 25 percent of the energy produced worldwide and 50 percent of the gasoline used by automobiles all over the world. The United States also uses 25 percent of the worldwide production of oil and gas. Each US citizen consumes 8 tons of oil equivalent (toe) per year. This is twice the amount used in Europe and 10 times the amount used in China.

US oil production, which represented nearly one-half of global production up to 1950, is no longer adequate to satisfy the country's demand. Starting in 1949, the United States was forced to import 10 percent of the oil that it consumed. This figure rose to nearly 50 percent in 1978 on the eve of the Iranian Revolution. Although the dependency ratio later fell to 30 percent during the 1986 price slump, it currently stands at over 65 percent. These oil imports exceed the output of the world's leading producer, Saudi Arabia, by more than 50 percent and represent more than 15 percent of global production.

In 2004, the main suppliers of crude oil and petroleum products to the United States were: Canada (15 percent), Mexico, Venezuela and Saudi Arabia (approximately 12 percent each), and Nigeria (8.5 percent).[6] OPEC currently provides 40 percent of all US oil imports. The leading oil exporters believe it is essential to have a presence in this huge market. The geographic proximity of Canada, Mexico and Venezuela makes them natural suppliers of the United States. However, for the Gulf countries, this desire for a presence in the US market comes at a cost, because crude oil shipped to the West is sold at a lower price than the same oil sent to Asian markets.

US gas production, which remained flat in the past few years, covers 84 percent of consumption. Neighboring Canada supplies 85 percent of import requirements, with the remainder provided by Trinidad and Tobago.

On the one hand, US demand for hydrocarbons is expected to increase in the coming years. On the other hand, US production will taper off, resulting inevitably in growing reliance on imports, with oil imports covering up to 80 percent of the country's demand. Canada is poised to be the perfect supplier on account of its proximity and the size of its reserves. In terms of oil, the development of the vast Canadian reserves of tar sands should allow the country to maintain its significant contribution to the US energy supply. However, Canada's reserves of natural gas (1 percent of the world's reserves versus 3 percent for the United States) will be insufficient to keep up with the region's increased consumption. Specifically, the growth in tar sands production will result in a further increase in gas consumption, since the heat necessary to exploit the tar sands will probably be produced by gas burning. North America will therefore become increasingly dependent on other countries for its gas supplies. The sharp rise in gas prices seen recently in North America has made it necessary for the United States to import natural gas in liquefied form. The existing terminals, built in the 1980s and barely used since then, are now being put to use and a number of new projects are being considered.

While Canada, Nigeria, Algeria and possibly Venezuela and Russia could offer partial solutions, the Middle East should become the principal source of US gas supply in future—with Qatar being a major supplier.

Qatar has vast reserves and is currently making significant investments to increase its production, which has already doubled in five years. The United Arab Emirates, which already has liquefaction plants, will also be among the top suppliers since Iran is still under a US embargo and Saudi Arabia has no export plans. Yemen and Oman are also expanding their export capacity.

European Union

The 25-member European Union consumes approximately 1.5 billion toe, or 15 percent of world energy consumption. Two-thirds of this energy comes from hydrocarbons (with oil constituting 44 percent and gas 24 percent). The continent's reserves are very limited. Norway is the only country with significant reserves. The European Union's ratio of proven reserves to production (R/P) is about 10 years for oil and 20 years for gas.

The measures taken during the 1970s – energy saving, development of the North Sea fields, use of nuclear power and development of renewable energy sources – allowed the European Union to reduce its oil dependency ratio from 97 percent in 1973 to 82 percent in 2004. With Norwegian production factored in, this ratio decreases to 64 percent for the entire European subcontinent. For gas, the corresponding figures for 2004 were 56 percent and 42 percent, respectively.

All the large oil-producing regions (Middle East, West Africa, North Africa, North Sea, Russia and South America) supply oil to the European Union. This situation stems from policies aimed at diversifying supply sources. The EU's main oil suppliers are Russia (25 percent) and Norway (22 percent). Europe imports 24 percent of its oil supply from the Middle East. In terms of gas, three countries (excluding the United Kingdom) are responsible for the bulk of European imports, namely Russia (20 percent), Norway (19 percent) and Algeria (12 percent). Given Europe's substantial needs, other suppliers have emerged, including Nigeria and Qatar, which supply liquefied natural gas to the European Union.

In the coming years, the European Union should be the region with the lowest rate of growth in energy demand—less than 1 percent per year. Nevertheless, given the depletion of the North Sea fields, oil imports will increase by nearly 50 percent and gas imports will triple. As a result, the EU's oil dependency could climb to 94 percent by 2030, its gas dependency to 81 percent and its total energy dependency to 70 percent.

Although imports from Russia, Norway and Algeria will continue to increase, those from the Middle East will represent the greatest increase in terms of absolute value. The transition from moderate to massive dependency will be instrumental in changing Europe's geo-strategic approach. The decline in European production, the uncertainty surrounding Russian supplies and the competition among the United States, China and India for a growing share of Africa's production of hydrocarbons should result in the Middle East playing a more prominent role.

Asia

Since 2002, Asia has consumed more energy than North America, accounting for nearly one-third of world consumption. Compared with the rest of the world, gas represents a smaller share of Asia's total energy consumption, ranking behind coal.

Overall, Asia imports only 10 percent of its gas supply thanks to surpluses in Indonesia and Malaysia (the two largest exporters of LNG in the world) and low demand in India and China. Imports from outside the region come almost exclusively from the Middle East, with Qatar in the lead, followed by Oman and the United Arab Emirates.

The situation is quite different for oil. The region depends on imports for 65 percent of its supply, 80 percent of which come from the Arabian Gulf. However, China has a much more diversified supply strategy than its neighbors.

Thus, the situations of individual countries are quite varied. On the one hand, the more developed countries (such as Japan, South Korea and

Taiwan), which have no resources, are entirely dependent on imports. China and India, on the other hand, are in the midst of an economic boom and have substantial energy reserves (mainly coal), as well as oil reserves, which allow them to meet 50 percent and 65 percent of their oil needs, respectively. In the next few years, however, increased Chinese demand for oil could represent nearly 50 percent of the total increase in demand worldwide. On account of declining production, China will probably need to import nearly 80 percent of its oil consumption by 2030. India is in a similar situation, with growing demand, production at a standstill and increased dependency on imports. In general, Asia's imports from the Middle East – which are already considerable – will only increase, primarily because of the continent's geographical position.

Currently, gas supplies from the Arabian Peninsula are not very significant. However, with the likely surge in demand in China and India, massive imports from outside the region will be necessary. Competition among China, Japan and the United States for access to Russia's oil and gas production marketed on the Pacific Rim will be intense, but dependence on (oil and gas) imports from the Middle East will remain very high.

Predominant Role of the Middle East

Concentration of Hydrocarbon Reserves

Proven worldwide reserves of conventional crude oil total approximately 157 billion tons, which is about 40 years of production at the current rate. Some 100 billion tons in additional reserves can be expected based on new discoveries and better knowledge of existing fields. In addition, the average recovery ratio, currently about 30 to 35 percent, could reach 40 to 50 percent in the future and further increase reserves. The contribution of these additional reserves will depend on oil prices and technological advances and will probably be spread out over time.

The distribution of the so-called conventional oil reserves is highly irregular with 62 percent being located in the Middle East—especially Saudi Arabia, Iraq, Iran, Kuwait and the United Arab Emirates. As a result, the R/P ratio of this region is approximately 80 years, while that of the rest of the world is barely more than 20 years.

There are also certain types of unconventional oil—those which cannot be produced using "conventional" methods. For example, the Orinoco basin in Venezuela is estimated to hold some 170 Gt (Gigatons or billion tons) of extra-heavy crude oil. With an estimated recovery ratio of 8 percent, the reserves in this basin would be 14 Gt using current technology while potential reserves are estimated at 40 Gt. These reserves would therefore exceed those of Saudi Arabia (36 Gt). However, this extra-heavy crude must be pretreated and transformed, using cracking processes, into lighter synthetic oil that can be easily transported and processed at a conventional refinery. In addition, tar sands, a mixture of sand and very heavy crude oil – bitumen or asphalt – are found in Canada (in Athabasca and on Melville Island). These resources are currently estimated at roughly 300 Gt, out of which 34 Gt may be recoverable.

Bituminous shales are rocks containing organic matter which can be partly transformed into hydrocarbons and which, when subjected to high heat, produce oil comparable to certain types of crude oil. Although they represent significant resources, production costs are quite high and current development techniques are very harmful to the environment.

Proven reserves of natural gas also total approximately 160 billion toe, for an R/P ratio of 67 years. As a result, natural gas could come to account for a much larger share of the world's total energy resources. Some 100 billion toe in additional reserves can be expected based on new discoveries. However, unlike oil, improvement in recovery ratios – which already average 80 percent – is hardly worth discussing. Unconventional gas resources appear to be considerable, but little is known about them.

Proven worldwide reserves of natural gas are concentrated in two main areas: the CIS (one-third) and the Middle East (41 percent). Three countries – Russia, Iran and Qatar – own nearly 60 percent of the world's gas reserves.

Strong Potential for Expanding Production

Energy resources are therefore sufficient to satisfy increased demand in the next 30 years. However, profound changes will occur in the geographic location of additional production capacity. After the intense diversification of the 1970s and 1980s, a new concentration of supply in the Middle East, as well as a few areas such as Russia, the Caspian Sea region and West Africa will materialize. Gas production, in particular, should develop in the Middle East, as the expected drop in transport costs will make it easier to ship gas to consumers in distant locations.

Despite the fact that it has more than 60 percent of reserves and the lowest production costs, the Middle East region currently produces only 30 percent and consumes only 7 percent of the world's oil. The Middle East's oil reserves could prove to be even higher than current estimates given the existence of very promising areas in Iraq and Saudi Arabia that have not yet been thoroughly explored. In terms of gas, the Middle East has 40 percent of reserves, produces 10 percent, and consumes 9 percent. This region therefore shows the greatest potential for development of hydrocarbon production. In view of the fairly moderate rise in demand forecast for the region, future increases in production could be used mainly to supply the world market and satisfy the overall increase in demand.

Oil production costs are low in the Middle East, being an average of US$3 per barrel, compared with US$15 for the most difficult fields. The same is true of investment costs. According to analysts, it would take US$6,000 to US$8,000 to increase production by one barrel per day in Saudi Arabia, whereas in most regions of the world this figure is at least three times as high. With crude oil priced at US$25, the pay-back period of the investment is under ten months, while with the current price (US$60) this period is less than four months. The production cost of non-associated gas is also lower in the Middle East than in Russia, North America or Europe.

Table 5.1
Hydrocarbon Development Potential by Region (2004)

	Oil					Gas				
	Reserves	Production	Consumption	R/P	Self-sufficiency ratio	Reserves	Production	Consumption	R/P	Self-sufficiency ratio
Middle East	62%	31%	7%	82	465%	40%	10%	9%	300	116%
CIS	10%	15%	5%	30	300%	32%	28%	22%	78	128%
Africa	9%	11%	3%	33	350%	8%	5%	3%	97	212%
South America	8%	6%	6%	41	143%	4%	5%	4%	55	109%
North America	5%	18%	30%	12	57%	4%	28%	29%	10	97%
Europe	2%	8%	19%	8	37%	4%	11%	19%	18	58%
Asia	3.5%	11%	30%	14	34%	8%	12%	14%	44	88%

Source: BP

The IEA predicts that, between 2000 and 2030, approximately three-fourths of the additional oil supply will come from the Middle East. The region would then represent 40 percent of the world's oil supply. Thus the challenge would be to increase Middle East production from 27 to 45 million barrels per day between 2005 and 2030, which implies building a total production capacity (additional capacity as well as replacement of depleted capacities) of more than 40 million barrels per day. Although this goal is clearly too ambitious to be attained, it points to the emerging challenges that need to be confronted.

Gas exports, on the other hand, should increase tenfold during the same period and the Middle East will become the top exporter of natural gas.

Need for Massive Investment in the Middle East

For the period from 2003 to 2030, the IEA forecasts an investment need of US$16 trillion, which amounts to US$568 billion per year, in the energy sector worldwide. More than one-half of these investments will be used to maintain existing facilities, while the other half will be used to satisfy the world's growing needs. More than 60 percent of these investments will be earmarked for electricity and 18 percent each for oil and gas.

An amount of US$100 billion per year will be invested in the oil sector, including 70 percent for the upstream sector (exploration and production), 20 percent for the downstream sector (refining and petrochemicals) and 10 percent for transport. Figures for natural gas are similar although a higher share will be invested in the gas transport sector.

The heaviest investments will have to be made in the high-consumption areas, with North America heading the list, followed by China. However, finding funds to build the necessary infrastructure in developing countries will be one of the major challenges. The investments are indeed enormous in relation to the size of the economies in developing countries and industrial projects in these countries are often more risky.

For all energy sources combined, the Middle East will – between 2000 and 2030 – need to invest US$1 trillion, including US$500 billion for oil and US$200 billion for gas. This is the area with the biggest investments in hydrocarbon production, given the size of the reserves and the urgent need to expand production to satisfy global energy requirements.

Stagnation in Production Capacity over the past 25 Years

Middle East crude oil production reached a peak of 22 million barrels per day (mbpd) in 1979 before dropping to 11 mbpd in 1985. This production drop was the result of the following factors:

- A decrease in oil demand following the tenfold increase in crude prices in the 1970s as a result of the oil crises
- The development of fields in areas such as Mexico, Alaska, the North Sea, Africa and Central America

Since that time, Middle East production has gradually returned to the levels of the 1970s, reaching nearly 25 million barrels per day in the year 2004.

Figure 5.1
OPEC Net Oil Export Revenues (1972–2006)

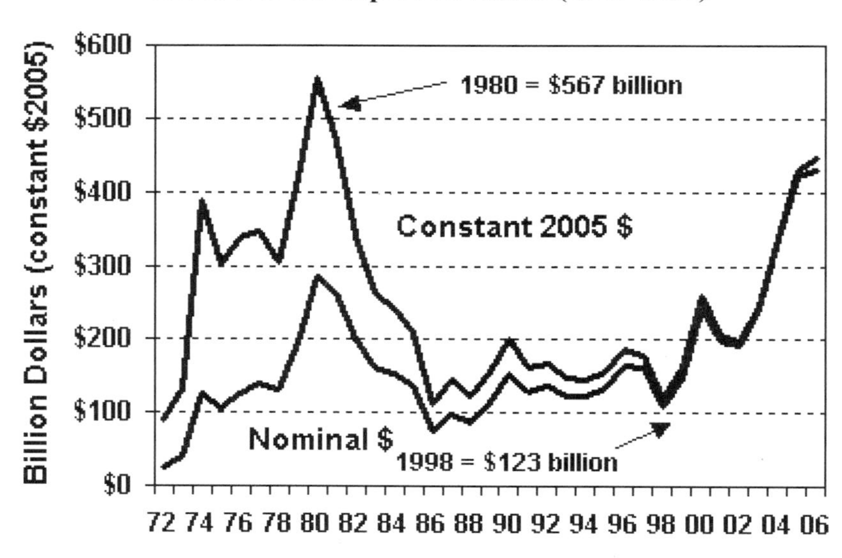

Source: Energy Information Administration, US Department of Energy (www.eia.doe.gov).

In the early 1980s, the Gulf countries thus found themselves with surplus production capacity, making investment unnecessary. The need to invest was felt only in the first years of the 21st century in response to higher-than-expected demand for oil, which caused a reduction in the region's excess production capacity.

There were other reasons for this under-investment. The decline in OPEC production in the 1980s, coupled with the price drop, was accompanied by a sharp decrease in the oil revenues of the member countries, just at a time when their populations were increasing dramatically. As a result, per capita income fell significantly. This in turn caused budgetary needs to skyrocket, leading to increased public debt and prompted governments to further tap into the finances of national companies, leaving them with too little capital to reinvest in the domestic oil industry. Moreover, since most Gulf countries did not take advantage of their earlier windfall oil earnings to diversify their economies adequately, they were forced to cope with high unemployment and had to implement costly social programs.

The countries with the largest reserves are also those with the lowest production costs. Economic logic would therefore have suggested that international investments in exploration and production be concentrated in these countries. In this case their production – through downward pressure on prices – would have squeezed out the development of more costly fields. However, to maintain high prices the OPEC countries, and especially those of the Middle East, reduced their output making possible the development of more costly fields in the North Sea, Alaska and new provinces like the Gulf of Mexico and West Africa. Ultimately, however, the swing producer is the producer with the lowest production costs—in this case, Saudi Arabia.

Table 5.2

Dependence of the Gulf Economies on Oil Revenues (2004)

Country	% of exports	% of GNP	% of budget
Saudi Arabia	90–95	40	70–80
Iran	80–90	—	40–50
United Arab Emirates	70	30	—
Kuwait	90–95	40	—

Source: Energy Information Administration, US Department of Energy.

In absolute value, the region's large producer countries – Saudi Arabia, Iraq, Iran, Kuwait and the United Arab Emirates – will supply most of the increased oil production, while the rise in gas production will primarily come from Qatar, Iran, Saudi Arabia, the United Arab Emirates, Oman and Yemen.

Saudi Arabia

Saudi Arabia has one-fourth of the world's oil reserves and produces one-eighth of the world supply, which makes it a key player in the oil sector.

For the past 25 years, it has played the role of the market's swing producer, adapting its output to supply. Traditionally, it has significant unused production capacity, although this has decreased from nearly 3 million to less than 2 million barrels per day.

Given the size of its reserves and the need to produce them over several decades, Saudi Arabia had for long maintained a "moderate" pricing policy aimed at preventing oil from being replaced by competing energy sources in consumer markets. Future price trends and the adequate physical supply of the market will depend on Saudi Arabia's willingness and ability to make the necessary investments in oil exploration and production in a timely manner. Production could be maintained at its current level simply by expanding the production of the 15 fields already in service. The Ghawar field alone, the largest in the world, produces 5 million barrels per day. According to Aramco's recently announced plans, national production should increase from 9.5 million barrels per day (mbpd) to 12 mbpd or even 15 mbpd by 2009. This production increase could be achieved by bringing on stream the 70 fields that have been discovered but not yet developed. In addition, the national oil company believes that its own cash flow will be enough to finance all the investments needed for such an increase in production.

Nevertheless, Saudi Arabia must deal with a number of constraints that are limiting its ability to invest in hydrocarbons. For example, the exceptional surplus budget revenues recorded in recent years (approximately US$25 billion in 2004) were allocated primarily to reducing the public debt, to funding education and development projects, to increasing the security budget in order to cope with terrorist threats (the budget for securing oil infrastructure totaled more than US$5 billion) and to providing direct assistance to the population. Indeed, unemployment has reached 15 percent and the population is growing at one of the fastest rates in the world (2.5 percent per year). This explains the decline in per capita oil revenues over the past 25 years which stands today at US$4,500, five times less than in 1980. Yet the economy remains largely closed and lacking in foreign investment.

In the natural gas sector, the "Gas Initiative," a major US$20 billion development project to produce non-associated gas over five years was under lively discussion for many years. This project was the first opportunity for foreign companies to invest money in Saudi Arabia since the nationalization of Aramco in the 1970s. The aim of this project was to integrate upstream gas activities with downstream activities (petrochemicals, electricity generation, desalination plants), thereby freeing up additional quantities of oil for export. In 2003, however, negotiations failed following disagreements regarding the return on investment demanded by international companies, headed by Exxon and Shell, and the size of the gas reserves to be included in the project. The project was then scaled down (to a few billion dollars), split into several projects and restricted to upstream gas activities. Lukoil, Sinopec, Shell, Total, ENI and Repsol–YPF are among the companies which will participate in these contracts.

Iraq

Iraq's oil production has always been low in relation to the size of its reserves, which are the third largest in the world, concentrated mainly in the country's central and southern regions. Since the war with Iran, the country has never returned to the production level it reached in the late 1970s. Exports resumed only after the introduction of the Oil-for-Food Program in 1996, under the auspices of the United Nations. Production is currently about 2 mbpd and is not expected to increase very significantly over the next few years, because of the political unrest. In the long term, however, the current production could be multiplied by three or four times.

The country's development potential is enormous, particularly as only 10 percent of its land area has been systematically explored. The western part of the country could prove to be one of the most promising areas in the world and could result in doubling the country's reserves. In addition, the investment cost is low—a few thousand dollars for the production of one barrel per day.

Although Iraq has enormous potential, its financing requirements are equally high. In the past 25 years, its infrastructure has sustained significant damage. In 2004 alone, the cost of the 642 attacks recorded by the authorities, particularly on the Kirkuk–Ceyhan pipeline, was US$10 billion in lost profits and repairs. Moreover, it appears that during the Saddam Hussein era, the shortage of new equipment and replacement parts as well as the lack of access to modern technologies may have damaged the fields.

Although the production, transport and shipping infrastructure would allow exports of more than 6 mbpd of crude oil, the closing of the export pipelines across Syria and Saudi Arabia (Saudi Arabia recently declared itself the owner of the IPSA pipeline that connects the two countries) and acts of sabotage limit exports to one-third of this capacity. The Kirkuk–Ceyhan pipeline and the three Iraqi export terminals on the Arabian Gulf are currently in operation (though often at reduced capacity). At present, the country's eight refineries are operating at a rate of only 50 to 75 percent, thus requiring it to import oil products. Therefore, dozens of projects are pending in the Iraqi oil sector.

Since most of Iraq's natural gas reserves are in the form of associated gas, an increase in the country's oil production capacity will, in itself, facilitate an increase in gas production. In fact, there is a gas pipeline project linking Kirkuk to Ceyhan. However, the country does not have a liquefied natural gas export terminal. In 2004, Iraq became part of the Arab gas pipeline project connecting Egypt, Jordan, Syria and Lebanon.

Iran

Iran has a very high potential for crude oil production, although actual output has not yet returned to the record level reached in the 1970s (currently 4 mbpd versus 6 mbpd at that time). An ambitious plan was recently launched in an effort to reach 5 mbpd by 2010 and 8 mbpd by

2020. Unlike other countries in the region, Iran has made major discoveries of new fields in recent years in Azadegan, Darkhovin, Dasht-e Adaban and elsewhere.

Nevertheless, a number of challenges need to be addressed, For example, the depletion rate of fields currently in production is significant, and infrastructures are nearing the end of their useful life. However, the National Iranian Oil Company (NIOC) lacks financial resources. The upstream sector is largely closed to foreign investors since the oil industry is supposed to be completely nationalized. A few foreign companies have entered into so-called "buy-back" contracts, which are considered to be services contracts and in principle, do not infringe on the government's sovereignty over hydrocarbon resources. It may be noted though, that these companies do not regard these contracts as very attractive.

Despite the high price of crude oil, Iran has chronic budget deficits, given the government's need to meet the requirements of a fast-growing population. Social assistance programs and subsidies to keep oil prices artificially low are burdensome and very costly. The implication is that the pump price is barely over distribution costs.

The embargo imposed by the US Congress in 1995 under the D'Amato Act continues to affect the country, with any company that invests more than US$20 million in Iran incurring US sanctions. Tension surrounding Iran's civil nuclear program and the election of Mahmoud Ahmadinejad as President do not suggest any normalization of relations with the United States in the near future. Of course, this situation does not encourage investments in the country's energy sector.

On account of its central location and access to the sea, Iran is the most natural channel for the outflow of hydrocarbons from the Caspian Sea region. Although the country has no production in this region, the development of swap agreements has allowed the country to free up quantities produced in its southern region for export purposes. Under these swap agreements, oil from the Caspian Sea is delivered to northern Iran to supply Tehran in particular, in exchange for crude oil produced in southern Iran that can be exported rather than sent to Tehran). However, these solutions will run into competition from the recently opened Baku–Tbilisi–Ceyhan (BTC) pipeline, and no large scale project will be possible as long as the US sanctions remain in place. Moreover, it will not be possible to develop production fully in the Caspian Sea region until its legal status has been resolved—an issue blocked by Iran so far.

The development potential of Iran's gas resources is enormous. Although it currently produces only 3 percent of the world's gas, the country has the second largest reserves in the world (more than 15 percent). Two-thirds of these reserves consist of non-associated gas and have not yet been developed. One-half of the energy consumed in Iran comes from natural gas and the government plans to make substantial investments aimed at increasing this percentage, thereby lowering its dependence on oil.

Development of the giant offshore South Pars field (containing 30–50 percent of the country's reserves), the Iranian portion of the Qatari North Field, is still underway. This is Iran's largest energy project, which, despite a number of delays, has already required more than US$15 billion. Sales from this field – for example, in the form of liquefied natural gas – could bring in approximately US$10 billion per year over 30 years. However, Iran for the moment has no liquefaction facilities and will have to face competition from regional countries already operating in this market (such as Qatar and the United Arab Emirates). Potential customers include Europe and Asia (India, Pakistan, China and others). Iran can also offer overland access to Asia's largest markets (China and India). Projects involving gas pipelines to India through Pakistan are currently under discussion.

Iranian President Ahmadinejad will most likely need large amounts of liquidity to keep his electoral promises, which seems possible only if hydrocarbon exports increase. Without greater openness to international companies, and the country's return to the fold of the international community (including. normalization of relations with the United States), there is little likelihood that the announced production targets will be met. However, development of the gas and possibly nuclear sectors gives the country several options for making larger quantities of crude oil available for export.

Qatar

Owing to its growing LNG exports and its low population levels, Qatar has not seen the same decline in GNP per capita as its neighbors. Recent budget surpluses were used to develop hydrocarbon production and transport infrastructure.

The country presents greater investment opportunities than its neighbors because it is as open to foreign companies as the nearby emirates of Dubai and Abu Dhabi. More than one-third of Qatar's crude oil production capacity is controlled by foreign operators.

Qatar has the world's largest gas reserves after Russia and Iran, and its strategy is to capture the Asian markets through long-term supply contracts. The first LNG delivery was made to Japan in 1996. The LNG export capacity of the Qatargas project (the country's first LNG export project) is expected to increase to 9 million tons per year in 2005 and RasGas-II's development should make it possible to supply 15 million tons per year to the United States by the end of the present decade. Exports to India, Spain and the United Kingdom are expanding. The largest gas-to-liquids (GTL) plant in the world (with a capacity of 140,000 barrels per day) is scheduled to start production around 2010.

The Dolphin Project, an integrated network of underwater gas pipelines between Qatar, the United Arab Emirates and Oman, is expected to cost US$10 billion and supply mainly gas-fired power plants in the Emirates, especially Dubai.

United Arab Emirates

As 94 percent of the country's oil reserves are concentrated in Abu Dhabi, this emirate plays the leading role in the UAE federation, created in 1971.

Meanwhile the production of crude oil from the emirate of Dubai has begun to decline. It may be noted that this oil serves as a benchmark in the Asian markets, and the drop in production undermines its status as a reference crude oil. Whereas Abu Dhabi joined OPEC in 1967, Dubai does not consider itself bound by the organization's quotas. The production potential is better utilized in the UAE than in neighboring countries because it has opened up to international companies. This allows the UAE to attract the investments needed for the development of the energy infrastructure.

The UAE government's efforts to diversify the economy have been successful, as evidenced by the fact that the country has become an important center for finance and regional tourism. The UAE is currently the only country – with Saudi Arabia – to have excess production capacity (approximately 250,000 barrels a day). The objective of the government is to increase the current production capacity of slightly more than 2.5 mbpd to 3 mbpd by the end of 2006.

Increased electricity consumption, greater needs in the petrochemicals sector and the size of the UAE reserves (considered the fifth largest in the world) are the main reasons for the investment of several billion dollars in the natural gas sector. The UAE is at the heart of the Dolphin Project, the cost of which is estimated at US$8–10 billion over 10 years. The project will begin with the construction of an underwater gas pipeline linking Qatar to Abu Dhabi. This pipeline will then be extended to Dubai and northern Oman. The aim is to satisfy Dubai's demand for gas, which is growing at a rate of 10 percent per year.

Kuwait

In 2004, Kuwait generated historic tax and budget surpluses. The country, which currently produces 2.5 mbpd, hopes to increase its capacity to 4 mbpd by 2020. As part of this effort, a US$7 billion, 25-year undertaking known as "Project Kuwait," launched in 1997, is considering the possibility of opening up a portion of the upstream oil sector to international companies. This measure would end a 30-year old taboo, attract investors and above all, demonstrate to international markets the government's willingness to contribute to the world's oil supply, precisely at a time when it has more tax revenue than ever and so little need for additional investments. However, these investments would take the form of "incentivized buy-back contracts," under which the government would maintain full ownership of the reserves and control over production levels, with companies as service providers, receiving payment for each barrel produced.

In addition, the country plans to spend US$5 billion between now and the end of this decade to modernize and increase its refining capacity. Owing to its refinery investments in other countries, Kuwait sells most of its oil in Europe on its own account. For example, it owns 8 percent of all Belgian service stations and is now showing similar interest in China and India. Lastly, the country plans to develop its petrochemicals sector, which would allow it to increase its revenues while adhering to OPEC quotas. Projects totaling more than US$1.5 billion have been launched in this sector.

Kuwait also plans to increase its gas consumption (mainly through imports from Qatar and Iran) in order to make additional quantities of oil available for export.

Inadequate Level of Current Investments

Clearly, there is enough money in the world to finance the energy projects analyzed in the foregoing section. However, the question is whether all the necessary projects will be completed. The main risk is that a lack of interest (for commercial, fiscal and other reasons) in new projects, rather than an overall lack of financing capacity, will limit investment. The energy sector must compete with other sectors of the economy to attract the necessary investments. Crude oil price trends will therefore play a key role. Another important factor will be the degree to which large producer countries open their hydrocarbon reserves to foreign investment.

The current situation is paradoxical. Never before have prices and profits been so high, and yet operators seem reluctant to make significant investments. On the one hand, production costs are only a few dollars per barrel in the Middle East, US$5–10 in the main production areas outside the Middle East, and US$15 in the most difficult areas. On the other hand, selling prices either equal or exceed US$50 and even touch US$60 per barrel. So why are the national oil companies of the producer countries (or the governments themselves) and the international private oil companies not investing more heavily?

Producer Countries

Despite the current high price, the large oil-producing countries prefer a cautious approach because of their historical experience. Producing countries in the Middle East and North Africa began to build modern economies in the 1970s and 1980s owing to substantial oil revenues resulting from the huge oil price increases of 1973–74 and 1979–80. Yet they saw their revenues plummet as oil prices fell to low levels (around US$18 per barrel) after 1986. They were forced to incur debt to finance budgets that had increased tremendously as a result of high operating expenses (equipment maintenance and health care, education and social assistance costs) and subsidies granted to a growing population. Consequently, the revenues of the national oil companies were heavily tapped to finance the state budget and the funds left over for exploration and production were then very limited. PEMEX – from a non-OPEC country but a very important producer – is a particularly good example of this. Mexico is an importer of gas although it presumably has abundant reserves. However, the revenues that PEMEX earns from hydrocarbon development are used primarily to fund the national budget at the expense of investment in the oil sector.

The reluctance of producer countries to open their upstream sector to foreign companies stems first, from their desire to maintain control of resources that are considered to be public property. Second, the OPEC quota system also makes it difficult for a producer country to attract foreign companies only to impose production limits on them afterward. Third, since these counties enjoy large budget surpluses, they have no incentive to call on foreign companies because they do not need more revenues at present. In Saudi Arabia, the failure of the Gas Initiative as mentioned earlier, demonstrated the limits of the desire for openness. Iran remains largely inaccessible to foreign companies because of international sanctions and unfavorable buy-back contracts. Iraq is still too unstable to make any projections. However, Kuwait has recently shown some encouraging signs. The United Arab Emirates and Qatar, both of which have continued to remain open to participation by foreign companies, are exceptions among the large producers in the Gulf region.

Another effect of the tendency to consider oil as a national asset that must benefit the entire population is that the prices of oil products are kept artificially low in domestic markets, at levels far below world prices. Sale prices are so low that the national company is not motivated to make significant investment. Saudi Arabia, for instance, has large amounts (more than several tens of billions of dollars) invested abroad in financial holdings – not necessarily in energy – whereas it could reinject a portion of that amount into the national economy.

Despite their growing power, the large producing countries, and especially those in the Middle East, are not interested in voluntarily limiting their production in order to maximize their immediate profits or in stockpiling reserves for future generations. Oil prices must remain reasonable to prevent a decline in demand. The economic and social stability of producer countries is tied to oil revenues, and it is in their interest to sustain high consumption levels of both liquid and solid hydrocarbons by their consumers for a long period of time.

The national oil companies of emerging countries (such as China, India and Malaysia) are highly dependent on the policies pursued by their governments. Chinese companies, with government backing, have adopted an aggressive foreign investment strategy. The main objective is neither profitability nor return on investment but control of reserves by a country with huge oil needs, as revealed by the CNOOC's unsuccessful attempt to buy out UNOCAL. In a less dramatic but very efficient manner, Petronas, Malaysia's national company maintains a strong presence in several producer countries, particularly in Africa.

International Oil Companies

The fall in prices in the 1990s led international oil companies to reduce their costs, very often through mergers and acquisitions. This led to cultural and structural changes in the industry, including the replacement of geologists with financial experts.

After a decade of under-investment, oil companies are ready to invest once again. However, it appears that they will limit their investments as long as the oil they seek is difficult and costly to develop.

The five largest international companies (Exxon–Mobil, Shell, BP, Chevron–Texaco and Total) posted more than US$80 billion in profits in 2004, but they are using a very large share of these profits to reduce their debt (which amounts to less than 10 percent of invested capital), buy back a portion of their shares and reward shareholders. Exxon-Mobil, which earned more than US$25 billion in 2004, bought back US$10 billion worth of its own shares to raise the share price, while Royal Dutch Shell made a commitment to distribute US$10 billion in dividends (on profits of US$18.5 billion). These companies are reluctant to invest for several reasons:

- *Lower barrel price expected*: They are expecting, in the medium term, a barrel price much lower than the current price. The consensus is US$25 to US$30 rather than US$50. They also believe that tax systems will be revised to skim off the surplus revenues. Whether this occurs under production-sharing contracts (through reduction of the amount of oil allocated to the foreign partner) or under concession agreements (where the government can collect a larger share of income by increasing royalty and tax rates), producer countries modify their tax schemes so as to increase their share of revenues in case of price increases and give foreign companies a fairly consistent portion (in dollars per barrel) of the income. This policy is in line with the predominant political approach, which views mining resources as the property of the people and the nation, the benefits of which must be reserved for the citizens of the producer country.

- *Unavailable reserves:* A large fraction of reserves, including those in the most promising basins, are not available to them. As discussed above, since the nationalizations of the 1970s, the OPEC countries remain generally reluctant to reopen their oil and gas sectors to large international companies. Outside the Middle East region, Venezuela has opened only its marginal fields and reserves of extra-heavy crude oil to foreign companies. Of the non-OPEC countries, Mexico remains totally closed to non-Mexican companies, and Russia, with the Yukos scandal, has shown that it wishes to maintain strict control over its reserves. This explains the repeated assertion by international companies that they "lack profitable projects."
- *Risky project conditions:* Where such projects do exist, oil companies often consider them too risky. Social instability in Saudi Arabia, the frequent attacks in Iraq and the uncertainty surrounding the future status of contracts, and Iran's international political situation are contexts that are not conducive to investment. The political stabilization of the oil-producing regions and the promotion of transparency, good governance and a clear regulatory framework are therefore necessary conditions for developing investments by international oil companies.

It can also be concluded that the trend toward deregulation and privatization seen in the high-consumption areas has had a somewhat negative impact on investment owing to the resulting emphasis on short-term profitability. Competitive bidding encourages investors to focus on the most profitable activities, at the expense of investments that lock up large amounts of capital, such as investment in infrastructure.

Bottleneck in the Refining Sector

The high price of crude oil and oil products has not yet become incentive to large investments either in exploration/production or in the refining sector. For nearly 20 years, the refining sector has been in a precarious financial situation (low profit margins and even significant financial losses) as a result of considerable excess capacity following the capacity expansion of the 1970s and the demand decline in the early 1980s. In 2005, worldwide refining capacity was barely more than it was in 1980. Substantial investments were made in Asia to satisfy the growing demand for oil products, but capacity has decreased in the western hemisphere. As in the upstream oil sector, the oil shocks of 1973–74 and 1979–80 led to widespread refining capacity surpluses. These surpluses in both the refining and production sectors were drastically reduced to very low levels in 2003–04 as a result of the sharp, unexpected increase in demand for oil products.

While the nationalization of oil fields transferred much of the exploration and production activity to the national oil companies of the producing countries, refining remained mostly under the control of private international companies, even though some national companies (for instance Aramco, PDVSA and KPC) developed a large refining industry in their own countries and invested in the refining sector in consumer countries (Aramco and PDVSA in the United States, KPC in Europe). Nevertheless, in the refining sector, more than in the production sector, the construction of new capacity depends first and foremost on private companies.

The strong reluctance of the populations of developed countries to allow construction of new industrial plants makes it difficult to imagine an increase any time soon in refining capacity that is consistent with the growing needs of consumers in large western countries. The "NIMBY" ("not in my backyard") syndrome hinders and even prevents the construction of new refineries. This increase must therefore occur either in the producer countries or in countries close to the consumer countries (the

Caribbean for the United States and North Africa, West Africa or the Middle East for Europe).

The refining sector also faces another challenge: available crude oils are increasingly heavy and rich in sulfur, while demand for oil products is increasingly focused on light fuels that are low in sulfur and high in quality. Refineries will therefore increasingly need to be equipped with sophisticated, costly facilities capable of transforming heavy oil fractions into light fractions.

Conclusion

Over the next 20 years, the oil-producing countries in the Arabian Gulf will have to make massive investments in production capacity for oil and gas, electricity and seawater desalinization. Major investments in the production and refining sectors may also be required, but those are less certain. Figures 5.2 and 5.3 below provide country-by-country details of planned upstream and downstream investments in the OPEC member states.

Figure 5.2

OPEC Upstream Development Plans (2005–2010)

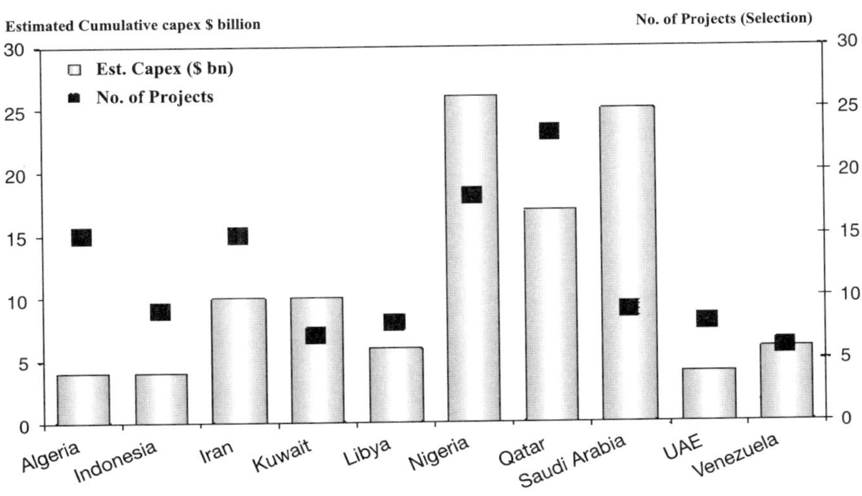

Note: Capex = Capital expenditure.
Source: OPEC.

Figure 5.3
OPEC Downstream Development Plans (2005-2011)

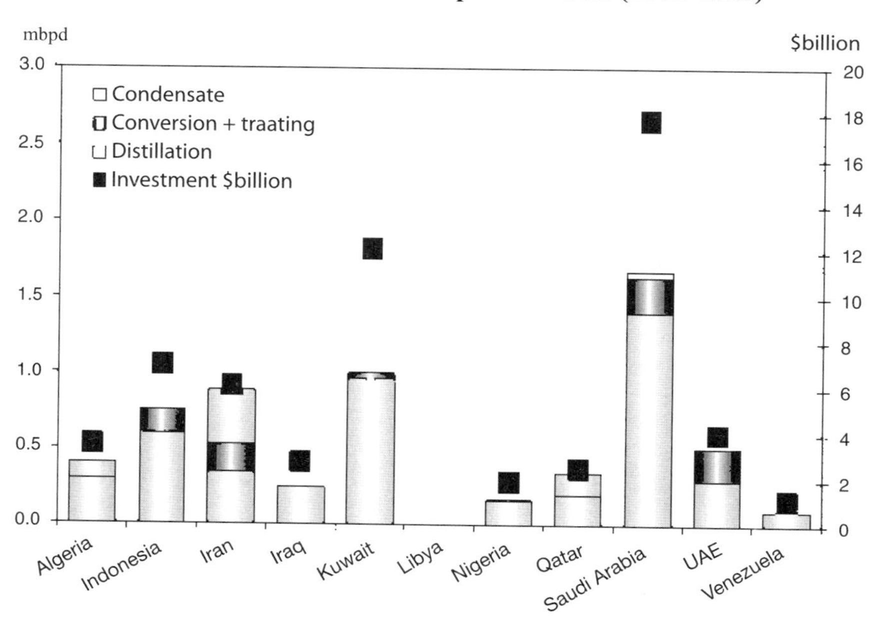

Investments in the refining sector forecast by the region's six OPEC countries total nearly $50 billion for 2005–2011. These investments correspond to an additional refining capacity of approximately five million barrels per day (mbpd), or a dozen new refineries. This would accommodate an increase of over 60 percent of existing capacity. The additional oil production would be devoted to meeting rapidly increasing consumption encouraged by extremely low consumer prices of petroleum products, and responding to Asia's continuing needs. Certain countries could enter into an association with foreign companies (for example, negotiations between Saudi Arabia and Total). New refining units could be set up to transform certain heavy and high-sulfur crude oils into light, low-sulfur crudes based on the model of heavy crude processing plants in Venezuela and tar sand processing plants in Canada.

However, the context in which these investments may be made is changing. For many years, OPEC and more specifically, the Gulf states, led by Saudi Arabia, has confirmed that it will supply the market with sufficient quantities at a reasonable price, thus guaranteeing supply security to consuming countries. To meet the high cost of investments, the possibility of opening doors to international companies was examined—an approach that could help to secure the necessary sums. Maintaining the price per barrel at around $70 changes the nature of the problem. Consumer countries are showing increasingly visible signs that they intend to reduce consumption or at least, limit its growth (for example, US president George Bush's speech and his statements regarding the diversification of sources and use of biofuels). Faced with the challenges of maintaining supply security, the producer countries are asking for "demand security," in other words, a guarantee that there will be a market for the oil produced as a result of their significant investments in production and refining. This position has triggered a series of chain reactions: high price of crude, stagnant demand, little investment incentive and high prices. International companies still want to invest in the Arabian Gulf, which holds the greatest hydrocarbon wealth. However, the situation is not particularly favorable for them. For more limited investments, national companies may prefer to choose service providers that are always ready to provide the necessary support.

Section 3

GULF OIL AND SUPPLY SECURITY

6

Saudi Arabia's Sustainable Capacity and Security Issues

Nawaf E. Obaid

This chapter offers an overview of Saudi Arabia's sustainable capacity, followed by a more specific focus on the security issues. These security issues have assumed greater importance in view of the systematic terrorist attacks in the Kingdom over the last two years, which have been attributed to the Al Qaida network. The following section highlights what Saudi Arabia represents to the Middle East region and the world today.

Saudia Arabia's Place in the Region and the World

The Kingdom of Saudi Arabia's strategic importance in both regional and global terms is summarized briefly below:

- *The center of the Muslim world*: As the birthplace of Islam and home to its two holiest cities, Saudi Arabia plays an unparalleled religious role for the world's Muslims.
- *The largest economy in the Middle East*: Saudi Arabia's GDP for 2004 was $254 billion, representing between 35–40% of the GDP of the entire Middle East (excluding Israel). The GDP estimates for 2005 showed that the Saudi economy would reach the level of approximately $330 billion.
- *The largest stock market in the region*: Saudi Arabia had an average market capitalization of $500 billion during the summer of 2005. This

represents around 45% of region's $1.1 trillion market capitalization. Saudi Arabia's stock exchange is the largest of the emerging markets, surpassing that of China, India and Russia.

- *Central to regional security:* The Kingdom has the largest and most modern military and internal security apparatus in the Gulf and continues to play an important role in the stability of GCC states.
- *The world's sole energy super-power*: Saudi Arabia is the largest oil producer and holder of largest oil reserves in the world. The Kingdom has 25% of the world's total proven reserves and a production capacity of 11 million barrels per day (mbpd) which is between 45–48% of total Middle East capacity. Saudi Arabia holds 40–42% of the proven reserves in Middle East.

In strategic terms, the first point is self-explanatory. The second point highlights the Kingdom's GDP figures, places them in the regional context and shows how it fits within the Middle East in general and within the Gulf region in particular. Saudi Arabia represents between 35–40% of the total GDP of the Middle East. The stock market is booming not only in the Kingdom but also in the United Arab Emirates and the other Gulf countries. The 2005 figures for Saudi Arabia actually represent just over 45% of the total market capitalization of the Arab world, and this astounding figure is rising phenomenally. The Saudi Stock Exchange (Tadawul) is actually the largest of the emerging markets, surpassing that of China, India and Russia. Another point to note is that the Kingdom is central to the region's security, being the largest and having the most modern military and internal security apparatus among the Gulf countries.

This brings us to the main theme of this chapter—Saudi Arabia's energy role and its predominant position in the global energy industry. In fact, as explained above, Saudi Arabia holds around 45–48% of the total Middle East production capacity and around 40–42% of the actual reserves in the Middle East. In order to delve into the subject at hand, it is necessary to examine Saudi Arabia's place in the energy market more closely. The following section pinpoints the factors that determine the country's place in the energy market both in global and regional terms.

Saudi Arabia's Place in the Energy Market

Saudi Arabia's predominant position in the energy market stems from several factors, the most important of which are listed below:

- *The largest oil reserves in the world*: Saudi Arabia claims 25% of the world's proven reserves (260 billion barrels), and 200 billion barrels more as a "possibility."
- *The largest oil producer in the world*: Saudi Arabia produces 12.5% of world total production, and has been the only oil producer that consistently sought to maintain surplus oil production in the past. The Energy Information Administration (EIA) forecasts that in 2025, Saudi production capacity will be 22.5 million barrels per day (mbpd).
- *An influential member of OPEC*: Saudi Arabia continues to play a central role in OPEC decisions due to its immense reserves and influence over the other member states, especially the Gulf countries.
- *Claims largest spare capacity*: The Kingdom has set a goal of maintaining 1.5–2.0 mbpd of spare capacity. It claims that it will have production capacity of 12.5 mbpd by 2009, and be "easily capable" of producing 15 mbpd within the next 15 years.

While the factors listed above are quite obvious, the figures deserve particular attention. The Kingdom of Saudi Arabia not only holds the largest oil reserves and is the largest oil producer in the world, but is also a highly influential member of OPEC. Moreover, Saudi Arabia's spare production capacity assumes great importance. Unfortunately, there is hardly any substantial spare capacity outside the Kingdom today, especially in such a tight market situation.

In Figure 6.1, Saudi Arabia's GDP for 2004 is compared to other countries in the Arab world, with the exception of Iran and Israel. It may be noted here that Iran would come in second and Israel third. Both are arguably neck and neck with the United Arab Emirates, which has had an unbelievable and impressive GDP growth over the last three years.

Figure 6.1
GDP Comparison for the Year 2004

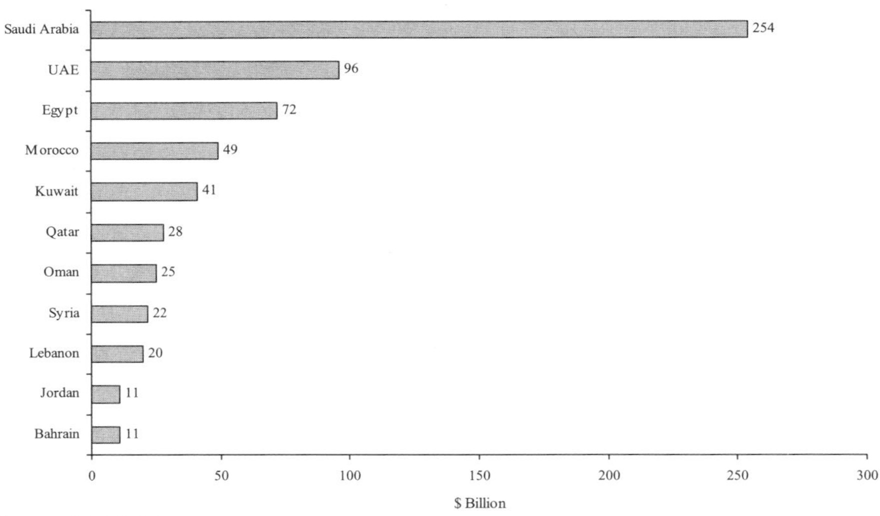

Source: Global Insight.

Figure 6.2
Saudi Oil Reserves
(billion barrels)

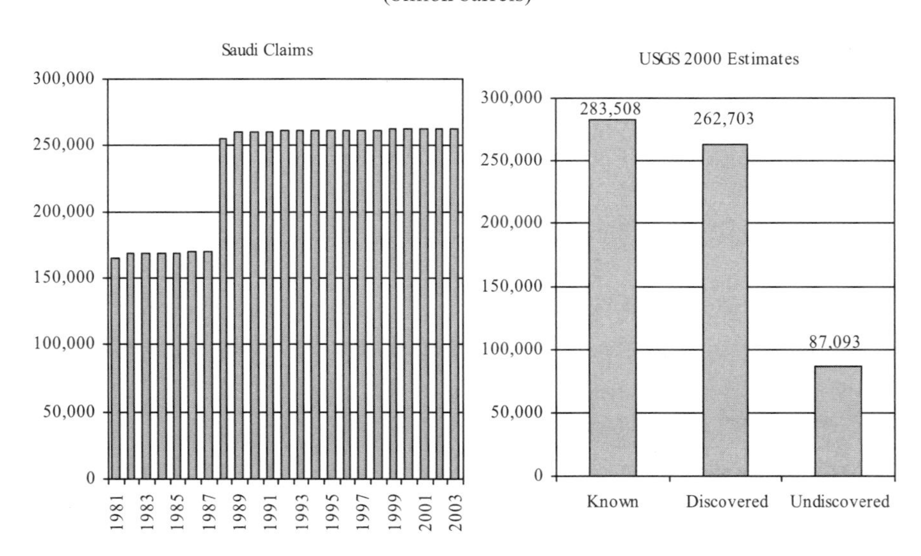

Saudi Arabian oil reserves (1981–2003) are depicted in Figure 6.2. These estimates are very straightforward and well known—just over 255 billion barrels, representing approximately 24–25% of the proven reserves in the world. The important issue at hand is what the Kingdom can do over the next five to six years to alleviate the supply constraints that are being witnessed today in the global oil market.

In this connection, it is worth quoting Ali Al-Naimi, the Saudi Minister of Petroleum and Mineral Resources, about Saudi Arabia's serious efforts to take steps towards a systematic decrease in the prices of crude over the foreseeable future.

Even before the Hurricane Katrina disaster occurred, Al-Naimi had announced, "To adequately fulfill this year's additional demand, Saudi Arabian output shall be increased from current levels at a later time this year."

After Hurricane Katrina struck the southern coast of the United States on August 29, 2005, Al-Naimi had declared:

> We are closely monitoring the impact of Hurricane Katrina on US crude oil supplies, refining activity and oil prices...we continue to be in close contact with our customers, especially those in the US, to assist them during any shortfall in oil supplies.

In the light of current market conditions and the destructive effects of Hurricane Katrina on US oil facilities in the Gulf of Mexico, Saudi Arabia is planning to increase production from 9.5 to 11 million barrels per day.

Figure 6.3 reveals the current status of Saudi Arabia's sustainable capacity. The year 2004 saw a small increase in Saudi Arabia's sustainable capacity as the new field from Qatif came online. Ever since March-April of the same year, there has been a comfortable sustained capacity of 11 million barrels per day, and that capacity has been ongoing all through 2005.

Figure 6.3
Saudi Production Capacity (2005)

(million barrels per day)

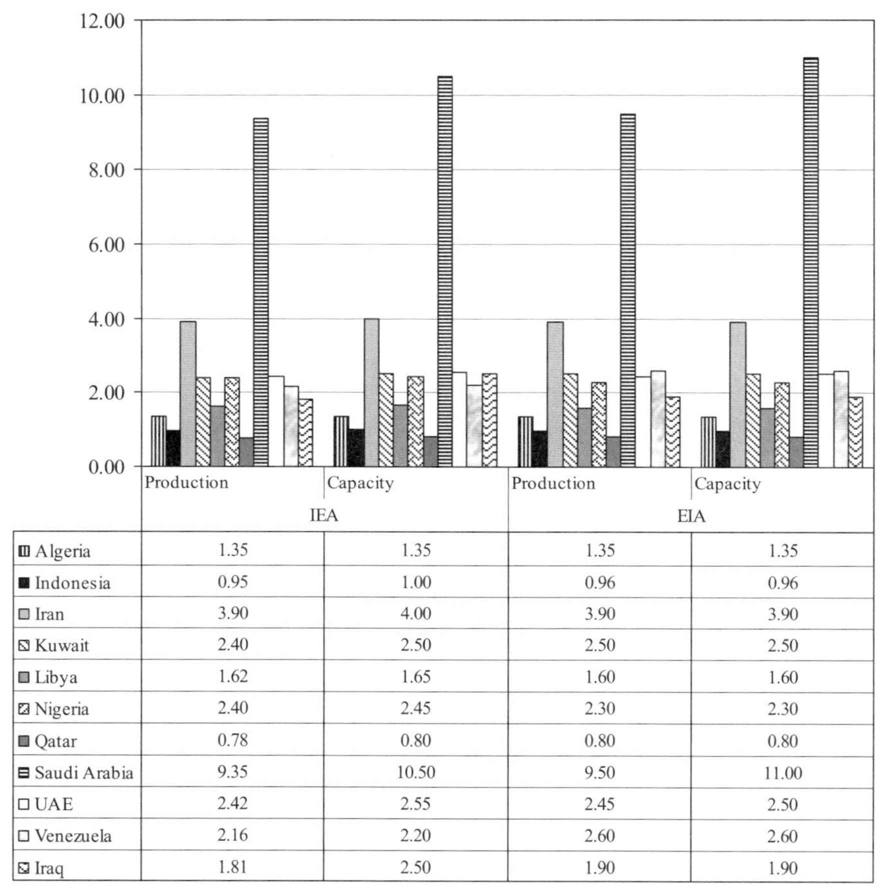

	IEA Production	IEA Capacity	EIA Production	EIA Capacity
Algeria	1.35	1.35	1.35	1.35
Indonesia	0.95	1.00	0.96	0.96
Iran	3.90	4.00	3.90	3.90
Kuwait	2.40	2.50	2.50	2.50
Libya	1.62	1.65	1.60	1.60
Nigeria	2.40	2.45	2.30	2.30
Qatar	0.78	0.80	0.80	0.80
Saudi Arabia	9.35	10.50	9.50	11.00
UAE	2.42	2.55	2.45	2.50
Venezuela	2.16	2.20	2.60	2.60
Iraq	1.81	2.50	1.90	1.90

Capacity (Saudi claims):

- 10.6 mmbpd in January 2004
- 10.8 mmbpd in February 2004
- 11.0 mmbpd in March 2004
- abandoned policy to eliminate excessive inventory buildup in the OECD on March 14, 2005

Production (Saudi claims):

- 9.25 mmbpd in February 2005
- 9.5 mmbpd in April 2005
- 9.6 mmbpd in August 2005

Note: Saudi capacity according to the IEA was 10.0-10.5 mmbpd, and the EIA was 10.5-11.0 mmbpd.

Source: Adapted from IEA *Oil Market Report,* April 12, 2005; and EIA *Short-Term Energy Outlook*, April 2005.

Figure 6.4
Saudi Oil Fields Production (2005)

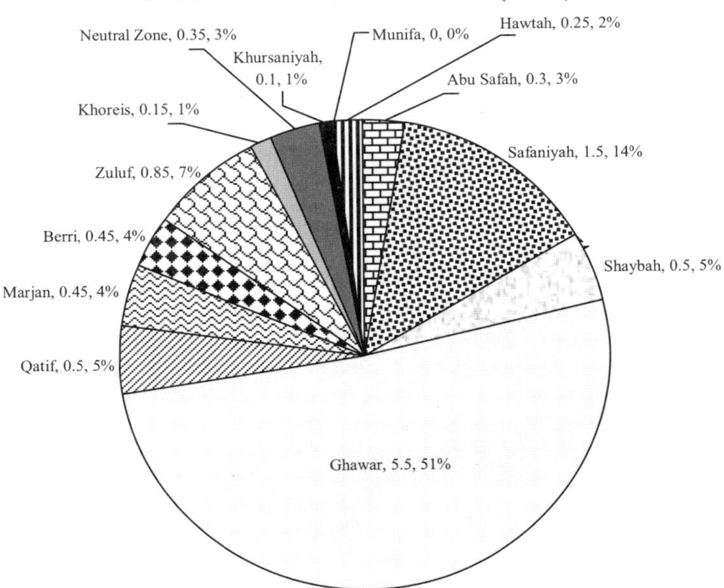

- Saudi Arabia has 80 fields and 1,000 wells.
- More than 50% of reserves are contained in eight fields.
- Ghawar and Safaniyah produce usually 65% of the Kingdom's oil.
- Munifa is offline.

Source: Saudi National Security Assessment Project

A detailed look at what Saudi Arabia is producing is provided in Figure 6.4. These are estimated figures, as the real figures are confidential. However, more or less, Saudi production would average anywhere from 9.5–9.6 million barrels per day. In Figure 6.4, the break-up of Saudi production in 2005 is provided on a field-wise basis. It may be noted that Ghawar produces approximately 50% of Saudi Arabia's global oil production capacity whereas the other fields contribute smaller amounts in percentage terms. Importance is also attached to the offshore Safaniyah oil field, which lies off the Eastern Province of Saudi Arabia. That is where most of the spare capacity is held in the heavier crudes, which, even though there are refining constraints can be exported to the countries that require it the most.

Figure 6.5
Oil Field Depletion Rates and Capacity

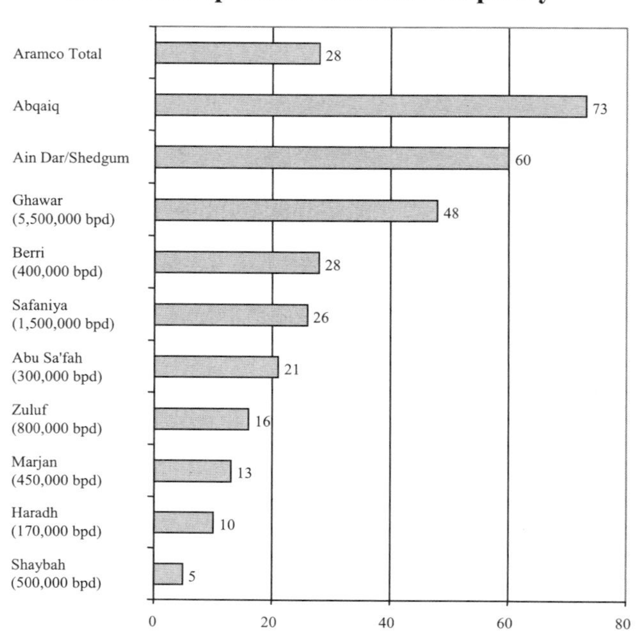

- Total depletion rate indicates the percentage of the estimated oil resources that has been pumped.
- Saudi oil fields total depletion rate is estimated to be 28%–30%
- To maintain the same capacity, more discoveries are needed.
- Total producable oil at given cost is very difficult to estimate, as is the gain from secondary and tertiary recovery.

Source: Saudi Aramco.

There is a lot of discussion about oil field depletion rates in the Kingdom, how much is being depleted and the location of fields that still need to be developed. Figure 6.5 offers a field-by-field illustration from Saudi Aramco, of how much oil has been produced and how much is still available. The current general rates of depletion in Saudi oil fields are approximately 28–30%. This point is important because of all the discussion in the press over the last couple of years regarding Saudi Arabia's spare production capacity and its actual oil resources.

Figure 6.6
Current Saudi Production Grade

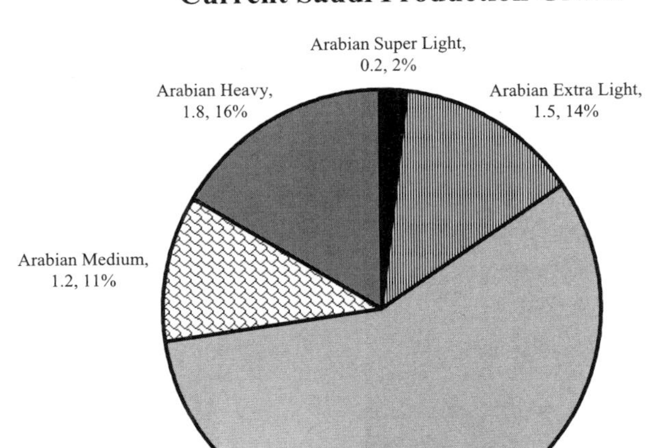

- About 70% of the Saudi production capacity is considered light gravity.
- The country is moving towards reducing the two heavier grades.
- Ghawar is the major producer of Arabian Light crude.
- Abqaiq is a producer of Arab Extra Light crude. An enormous field, it contains 17 billion barrels of proven reserves.
- Shaybah has estimated (EIA) reserves of 15 billion barrels. It produces a mix of Arabian Light and Arabian Extra Light.
- Munifa is still offline, but it could reach 1 million bpd of Arabian Heavy.

Source: Saudi National Security Assessment Project

Figure 6.6 focuses on current Saudi production grades and provides details of the different grades currently available in the Saudi production quota. As indicated, approximately about 70% of the Saudi production capacity today is considered light gravity, which is suitable for the existing plants. Fortunately, Ghawar, which is the largest producer of Arabian Light crude, actually produces the lighter version, so it is very affordable and considered suitable for the refineries. The point is that most of the Saudi's production sustainable capacity will be concentrated on the extra light crude in order to avoid the current situation where oil is available, but unfortunately, cannot be extracted or exported to the big markets because of the refining capacity problem.

Table 6.1
New Saudi Production
(contributing to 12.5 million bpd capacity in 2009)

Oil Field	Grade	New Capacity (bpd)	Date
Abu Safah & Qatif	Arab Light & Extra Light	500,000–550,000	2004/2005
Haradh	Arab Light	*300,000*	2006
Khursaniyah	Arab Light & Extra Light	*500,000*	2007
Shaybah	Arab Extra Light	*300,000–400,000*	2008
Nuayyim	Arab Extra Light	*100,000*	2009
Khoreis	Arab Extra Light	*1.0–1.2 million*	2009
Total		**2.70–3.05 million**	**2004–2009**

Estimated sustainable capacity in March 2005	11,000,000 bpd
Estimated increase in capacity on stream 2006-09	+ *2,400,000 bpd*
Will replenish the natural decline 2005-09	– 800,000 bpd
Estimated sustainable capacity in 2009	= 12,600,000 bpd

Note: this is not $3 per incremental barrel oil. Investment costs will reach $18 to $20 billion between 2003 and 2009.
Source: Saudi National Security Assessment Project

The Saudi production program, which aims to increase sustainable capacity over five years from 2004 to 2009, is detailed in Table 6.1. Unfortunately, this program will bring on stream most of the new incremental net increases in oil production not only with respect to the region of Saudi Arabia and the Gulf but also the world as a whole. This is why it assumes importance. Over this period, Haradh, Khursaniyah, Shaybah, Nuayyim and Khoreis are the five fields that would bring the incremental increase to the final tally of 12.5–12.6 million barrels of sustained capacity in 2009. Because of the huge increases in the energy budget, coupled with the situation in the oil markets, there is an expectation that this figure will be met by mid-2008. Unfortunately, except for Kuwait and the United Arab Emirates, no other country so far actually has a program in place to increase its net capacity. So, over the next two to three years, there will be only small incremental increases in capacity. For Saudi Arabia, this target will be met hopefully by mid-2008—a year that will witness a big increase in Saudi Arabia's sustainable production capacity.

Table 6.1 also gives an indication of the estimated investment costs involved in adding this new capacity. This amount is expected to be US$18 to US$20 billion just on the upstream sector. The downstream and refining sectors will be discussed later. However, US$20 billion is only an estimate of what it will cost and this is expected to increase necessarily as attempts are made to bring the fields online.

Mega Projects to Develop Saudi Fields

Saudi Arabia has undertaken some mega projects to develop particular fields and bring them on stream. Summarized below is the status of each of these projects, which include Munifa, Abu Safah and Khatif, Kursaniyah, Khoreis, Haradh and Shaybah:

Status of Mega Projects

1- Munifa:

- In January 2004 the field was offline
- Aramco claims it could produce up to 1 million bpd in the foreseeable future (no decision has been made to develop the field because it produces Arabian Heavy).

2- Abu Safah & Qatif:

- Completed in late 2004 at a cost of $4 billion
- In January 2004 these fields produced 300,000 bpd
- In 2005 these projects came on stream and produced 500,000 bpd.

3- Khursaniyah:

- Signed in March 2005. The budget was approved for $4 billion (up $1 billion from initial estimates)
- In January 2004 this field produced 100,000 bpd
- By 2007 Aramco claims production will reach 500,000 bpd.

4- Khoreis:

- The budget was approved for $6 billion (up $1 billion from initial estimates)
- In January 2004 this field produced 150,000 bpd
- By 2009 Aramco claims production will rise to 1.0–1.2 million bpd.

5- Haradh:

- Inaugurated in January 2004, its estimated cost is $1 billion
- In January 2004 the field was reported to produced 170,000 bpd
- By 2006 the project will expand production to 300,000 bpd.

6- Shaybah:

- Its estimated cost is $1.5 billion
- In January 2004 the field produced 500,000 bpd
- By 2008 Aramco claims its new capacity will be 300,000–400,000 bpd.

It is worth noting here that according to Aramco, all of these fields (except Munifa) will be developed.

All these field developments are called mega projects because their cost by field is huge. Abu Safah and Qatif are the last of the major fields that came on stream in 2004 at a cost of US$4 billion. The other fields, which will be productive within the next 1–3 years, all have rather similar costs. Khursaniyah is the next project that will add to Saudi production increase. Initially estimated to cost US$2.5–4 billion, it has now reached US$5 billion. The biggest of all is Khoreis, which will represent most of the incremental production. The Khoreis project was initially estimated at US$6 billion and its budget has now reached US$7 billion with the cost of bringing all the rigs and equipment needed to speed up the process. Haradh is expected to cost US$1 billion, which is small in comparison to the overall cost. Moreover, Shaybah, which had already cost US$4 billion to become productive, will cost an additional US$1.5 billion to increase its capacity. Finally, it may be noted that Shaybah produced 500,000 barrels per day in January 2004, and after this first level of increase in Saudi production, it will average around 200,000 barrels per day.

Downstream Activities

Finally, it is important to know what is the new refining capacity planned by Saudi Aramco. The company wants to bring online whatever is feasible. If Munifa, which is in the heavier grade, is made operational, this would by itself add another 1 million barrels per day. That project would form part of Phase Two, scheduled to be completed by 2009–2010.

It is necessary to pay attention to downstream activities, which is a problem area in the energy sector, and particularly to the downstream refining portion of the petroleum expansion program. Figure 6.7 provides details of current Saudi refining capacity and its intended expansion.

Figure 6.7
Saudi Refining Capacity

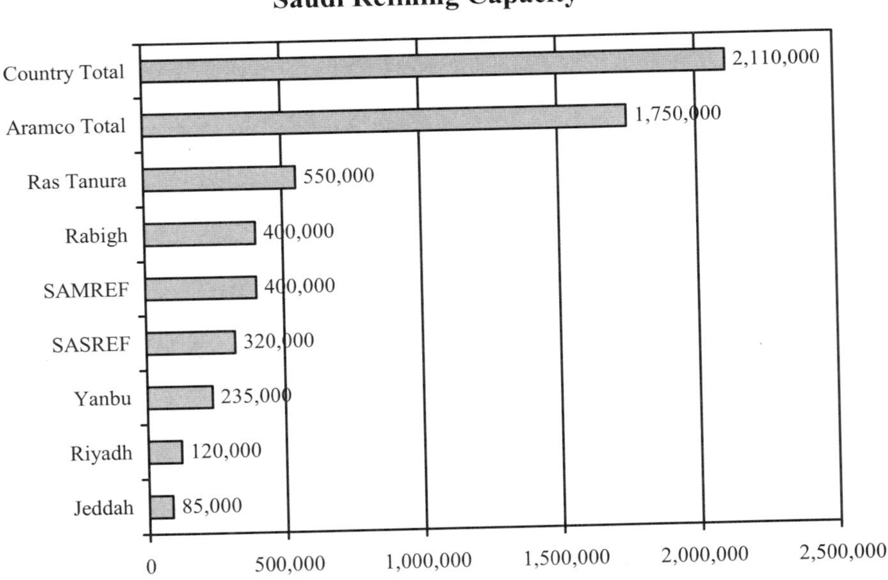

- Saudi Arabia has 8 refineries, with a combined crude throughput capacity of roughly 2.1 million bpd, and about 1.75 million bpd of overseas refining capacity.
- The Kingdom plans to upgrade and expand the Rabigh refinery by 425,000 bpd.
- The Kingdom has plans to expand its refining capacity in North America, to build 2 new domestic and 3 new overseas refineries in next 5 years.

Source: Saudi Aramco.

Saudi Arabia has eight refineries plus one in the Neutral Zone, with roughly 2.1 million barrels of refining capacity per day and an extra of 1.75 million barrels per day in outside joint ventures with big international oil companies. The big plan for the downstream is the expansion of the Rabigh refinery, which will add another 425,000 barrels per day to bring it to some 825,000 barrels per day. More broadly, the Kingdom has plans to expand its refining capacity in North America, to build two new domestic and three new overseas refineries, of which two are said to be in the planning stage.

Saudi Refining Expansion Program

The Saudi Refining Expansion Program proposes to increase Saudi refining capacity from about 3.9 million to over 6.0 million bpd and includes the following projects:

- *MOTIVA Enterprises (Texas)*: Expansion is intended to increase capacity from 235,000 to 600,000 bpd (Cost: $4 to $5 billion, with Aramco owning 50%).
- *Fujian Refinery Project (China)*: With SINOPEC and EXXONMOBIL, this project plans to increase capacity to 230,000 bpd (Cost: $3.5 to $4 billion, with Aramco owning 25%).
- *Qingdao Refinery Project (China)*: With SINOPEC, this project is set to add 200,000 bpd (Cost: $1.2 to $1.5 billion, with Aramco owning 25%).
- *Rabigh Refinery Expansion*: With SUMITOMO, the plan is to increase capacity from 400,000 to 825,000 bpd (Cost: $6 to $7 billion, with Aramco owning 50%).
- *Yanbu Refinery New Project*: This project will adds 425,000 bpd with an international partner yet to be selected (Cost: $4 to $5 billion, with Aramco owning 50%).
- *Jubail Refinery Expansion*: Plans for 400,000 to 450,000 bpd capacity project with an international joint venture partner (Cost: $4 to $5 billion).

The Saudi refining expansion program should bring the current capacity from 3.9 million barrels per day to a target of approximately 6 million bpd, hopefully by 2011–2012. Obviously, the costs involved in these projects are considerable. With the heavy price tags mentioned earlier for the upstream sector, all these sizeable expansion programs will entail a massive budget.

US-Saudi Special Relationship

From a more general perspective that puts the oil issue into a geopolitical context, the special relationship between the United States and Saudi Arabia needs to be understood. The Abdullah–Bush Summit in Crawford, Texas in April 2005 highlighted developments in the energy sector, and also indicated how Saudi Arabian policy would proceed over the next five years.

- The US "signaled" to the Saudis that it wanted a commitment to increase oil supply in the short-run to ease high oil prices.
- Saudi Arabia reiterated its view that one of the reasons for the high gas prices is the bottleneck created by aging US refineries and indicated that Saudi investments in US refineries were possible.
- The Saudis presented their plan to invest $50 billion in the energy sector to increase production capacity to 12.5 million bpd by 2009 and to reach 15 million bpd within 15–20 years.
- The Saudi Petroleum and Mineral Resources Minister Ali Al-Naimi stated that the Kingdom would increase its petroleum production in the light of Hurricanes Katrina and Rita.

The joint statement issued by President Bush and Saudi Crown Prince Abdullah indicated the importance of the ties between the United States and Saudi Arabia. It stated:

> Both nations pledge to continue their cooperation so that the oil supply from Saudi Arabia will be available and secure. The United States appreciates Saudi Arabia's strong commitment to accelerating investment and expanding its production capacity to help provide stability and adequately supply the market.

So, as H.R.H. Prince Saud Al-Faisal said in his presentation at the Baker Institute in September 2005, two new refineries are being planned in the Kingdom. The initial investment in the energy sector will be US$50 billion on average over the next five years. This amount takes into the account not only the upstream and the downstream but also the existing fund pledges for the expansion of Saudi Aramco's production over the next five years. Collectively, these investments could lead to a final sustained capacity of 12.5 million barrels per day by 2009.

Finally, Saudi Arabia must focus on concerns about security and stability. This issue has come to the fore, especially over the last three years, due to the Al Qaida terrorist bombings in the Kingdom. It may be noted that basically, there is a serious weakness in the Saudi economic structure. This is true not only of Saudi Arabia but also the other Gulf states. Oil revenues made up 90–95% of the total Saudi export earnings, comprised 70–80% of the state revenue, and approximately 40% of the Kingdom's GDP in 2004/5. So it is evident that the required economic diversification has not actually taken place.

Oil Revenue-Funded Stimulus Package

In view of the high oil revenues that the Kingdom has enjoyed in the past two years, there has been a sizeable revenue surplus that has been used to fund a stimulus package, which was considered necessary in addition to existing projects in the public policy sectors. The budget surplus has provided an opportunity for the leadership to practice transparency in government spending, support social programs and entitlements and improve aging infrastructure.

The following are the specific allocations under the oil-revenue financed stimulus package totaling $32 billion for 2005–2006:

1. *Salaries:* $8 billion to increase the salaries of government employees (15% raise)
2. *Services and Infrastructure:* $10 billion allocated for development and maintenance of services and infrastructure, including:

- $2.13 billion for the building of public housing projects
- $1.86 billion for construction of new desalination plants
- $1.33 billion for construction of new highways and roads
- $1.2 billion for street maintenance and drainage system
- $1.06 billion for construction of new schools
- $1 billion for the construction of university campus construction
- $800 million for construction of primary health care facilities
- $666 million for construction of new vocational training institutes

3. *Exports:* $4 billion allocated for the Saudi Export Program Initiative
4. *Industrial Development:* $3.46 billion to increase the capital of the Saudi Industrial Development Fund
5. *Real Estate:* $1.2 billion to increase the capital of the Saudi Real Estate Fund
6. *Credit:* $800 million to increase the capital of the Saudi Credit Bank
7. *Social Security:* $4 billion to increase the minimum social security payment.

Overall, the stimulus package amounts to US$32 billion, of which the Saudi government will spend US$5.33 billion on repaying its own domestic debt. More specifically, it has funded an increase in the military and security budget, which amounts to approximately US$10 billion today. In addition, the budget surplus helps to tackles more structural problems, by funding social programs and entitlements intended to bring down the high unemployment rate, and by financing the revamping or replacement of aging infrastructure. Perhaps the most important aspect is the amount of US$8 billion earmarked to raise the salary level of Saudi government employees. This is not just a one-time increase in budget but is meant to go forward for all the years to come. Thus, all government employees in the Saudi administration will have a 15% salary increase.

The stimulus package also makes specific allocations for existing social and public policy programs. For the first time US$4 billion is being allocated for the Saudi Export Program Initiative, which aims to make the country's exports more competitive to the outside world. Again, more funds have been allocated to augment the capital of the main Saudi borrowing institutions, whether at home or abroad—the Saudi Industrial Development Fund, the Saudi Real Estate Fund and the Saudi Credit Bank. This includes increasing the capital for the minimum social security payment for people who are on welfare.

Oil Infrastructure Security Issues

The issue of oil infrastructure security must be addressed adequately. The Saudi security budget, which stood at over US$8 billion in 2004, has generally increased and over the last three years it has averaged between US$8–US$10 billion, of which a small identifiable portion, in comparison to the regional budgets, has been earmarked specifically for its oil installation security. In addition, an average of between US$1.2–1.5 billion per year is spent on the various security-related services provided by different ministries in Saudi Arabia, including the Ministry of Defense, Ministry of Interior and the National Guard. This spending is consistent with the need to safeguard the country's oil installations.

At present, there is air surveillance of oil installations from helicopters as well as round-the-clock F15 patrols. On the perimeter of these installations, heavily equipped National Guard battalions stand guard. At any given time, it is estimated that there are between 25,000 and 30,000 troops protecting the most vital oil infrastructure. These include troops provided by different ministries. Saudi Arabia's oil terminals are similarly well defended. Each terminal and platform has its own specialized security units, comprising Saudi Aramco security forces, specialized units of the National Guard and the Ministry of Interior's petroleum installation security force. Additionally, the Coast Guard and components of the Navy protect the oil installations from the sea. All this goes to show how seriously the Saudi government takes the defense of its oil installations.

However, there are some general vulnerabilities that need to be considered. If an attack occurred, the whole infrastructure would not necessarily collapse. Obviously, there are certain vulnerabilities in the oil infrastructure, such as the 17,850 kilometers of pipeline system within the Kingdom. Although extra pipelines have been stored approximately every 300 to 400 km, in order to repair any damage within 24 to 48 hours, yet the possibility of an attack must be taken into account.

Under the Ministry of Interior (MOI), various security services are involved in safeguarding petroleum infrastructure: Special Security Forces, Special Emergency Forces, the General Security Service, regular forces of the Public Security Administration, the Petroleum Installation Security Force (PISF), specialized brigades of the National Guard, the Navy and Coast Guard. However, it is important to note the conclusions of a recent assessment by one of the major security ministries in Saudi Arabia. Its finding was that, barring a spectacular strike on the scale of 9/11, or some form of systematic sabotage from inside Saudi Aramco or other key energy industries, (which, incidentally, is now highly unlikely) most foreseeable assaults are likely to be confined quickly and any resulting damage repaired relatively soon. This is based on simulation programs carried out by Saudi Aramco and one of the ministries on the potential for attacks, their likely locations and probable impact. Nevertheless, energy security will continue to pose a problem for Saudi Arabia whether it is an internal attack staged from within Saudi Arabia or launched from outside. Moreover, with global energy use expected to rise more than 50% by 2025, the security of Saudi energy exports will play an increasingly vital role in the world economy.

Measures for Economic Security and Stability

Finally, it is necessary to pinpoint the measures the country must undertake to ensure its economic security and stability. This is not just a physical issue but also a question of reform and development. What is clear is that, notwithstanding its incoming oil revenues Saudi Arabia, on account of its centrality in the Middle East and its importance to the global economy, must ultimately embrace diversification, which is the only way to sustain the high oil prices. Without suitable diversification to sustain high oil prices, Saudi Arabia would not be able to fuel the huge economic boom and growth required to meet domestic needs. Therefore, some social reforms are at least as important as political reforms. Despite many of the positive steps that the Kingdom has taken in the last three years, much remains to be done.

Apart from diversification, which is critical, there are the problems of demographics and youth explosion. In late 2005, a sizeable company program has been put into place, and is being implemented very seriously. Saudization is necessary to deal with the high unemployment rate in the Kingdom. There is also the issue of some social and political change. The privatization campaign is also very important. It was moving slowly in the past and was unable to build a robust private sector. However privatization is gaining momentum and is a major feature that will allow the Saudi economy to diversify in areas away from the oil sector.

Finally, a realistic and concerted effort has to be exerted in dealing with the problems of demographics and unemployment to limit the potential pool of recruitment by extremists. This is a significant issue with regard to the Al Qaida network, which has already been dealt with to a large extent. Unfortunately, there is still a very small pool of terrorists in the Kingdom who might pose a threat both inside and outside Saudi Arabia, and the government is doing its utmost to deal with such groups.

7

The Future Importance of the Gulf in US Oil and Gas Requirements

Aloulou Fawzi

According to the Energy Information Administration's (EIA) annual outlook for energy markets through 2025, the oil import dependence of the United States is forecast to grow from 56 percent to 68 percent in the reference case. The market share of the Gulf region in US oil imports is forecast to grow from 20.4 percent in 2003 to 30 percent in 2025. The EIA's outlook is based on the *Annual Energy Outlook 2005* (AEO) published in February 2005[1] and the *International Energy Outlook 2005* (IEO) published in July 2005[2] (See Figure 7.1).

The Gulf region in this case includes Kuwait, Qatar, Saudi Arabia and the United Arab Emirates plus Iraq and Iran.

Figure 7.1
Petroleum Imports into the United States by Source (2003 and 2025)
(million barrels per day)

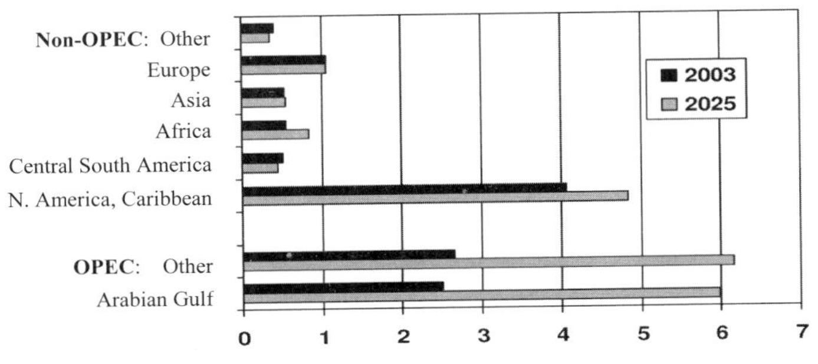

Source: *Annual Energy Outlook, 2005.*

The future growth in US natural gas supplies is also projected to depend on more import from the Arabian Gulf mainly through liquefied natural gas (LNG) imports. About 50 percent of the US imports of natural gas are projected to be in the form of LNG by 2010 and 40 percent of this could come from the Gulf region.

It is important to note that because of the uncertainty in the world oil markets, the EIA's most recent outlook for oil capacity through 2025 has been revised downward in the Arabian Gulf region and mainly in Saudi Arabia. The latest EIA forecast in the reference case projects that Saudi Arabia will produce 16.3 million barrels per day, and the Gulf will produce only 39.3 million barrels per day by 2025. This would mean the Gulf region's production would be 5.7 million barrels per day less than the previous forecasts (See Figure 7.2).

Figure 7.2
Gulf Oil Productive Capacity by Country (2002 and 2025)
(million barrels per day)

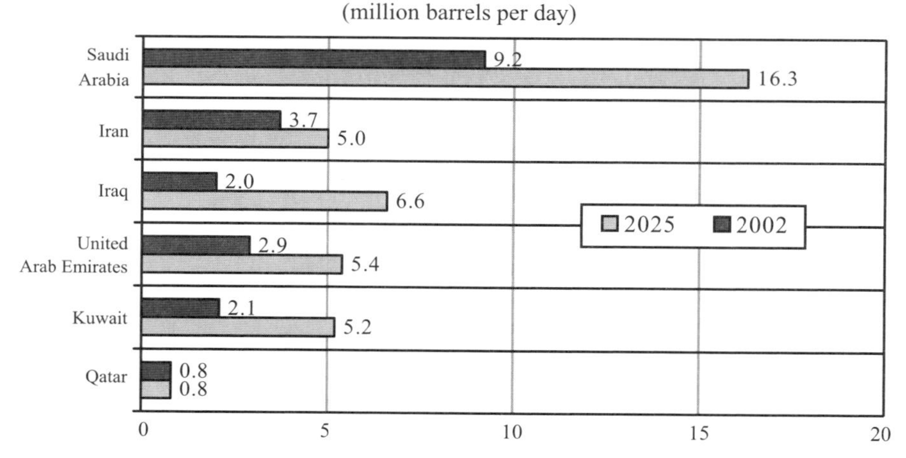

Source: *International Energy Outlook, 2005.*

A detailed analysis of the US energy markets by fuel and by sector will explain the dynamics behind the growing dependence on oil and gas imports and the future importance of the Arabian Gulf in US energy requirements, based on a reference case where the price of oil is projected to reach US$30.31 per barrel (in 2003 US dollars) by 2025, and a higher oil price scenario called "High B" where oil price is projected to reach US$48 per barrel (in 2003 US dollars) by 2025 (See Figure 7.3).

Figure 7.3
Oil Prices (1970–2025): US Refiners' Cost of Imported Oil per Barrel
(in 2003 US dollars)

Source: *Annual Energy Outlook*, 2005.

The EIA 2005 forecast of world oil markets in 2025 was developed based on a linear programming model that balanced oil supply with demand. The demand side is driven by gross domestic product (GDP) and population growth for all regions of the world. The supply side is determined by an estimate of the world oil resource base that could be produced under different oil price cases. The EIA models first determine how much non-OPEC oil (both conventional and non-conventional) could be produced at the assumed price path. An important criterion was whether producers would receive an adequate rate of return on their investment (usually 10 percent). With total non-OPEC supply having been established, the second step assumed that the remainder of the worldwide demand would be met by OPEC producers and determined an appropriate production capacity for each OPEC producer. In other words, the assumption is that all non-OPEC oil projects that show favorable rate of return on investment would be funded, and that OPEC would act as the residual supplier.

US Energy Mix Outlook: An Overview

In the United States, the total energy use through 2025 will grow at an average annual rate of 1.4 percent, slightly less than half the projected rate of economic growth estimated at 3.1 percent. Fossil fuels – oil, coal and natural gas – are expected to account for an increasing share of total energy use. The petroleum demand is driven by growth for transportation uses while demand for natural gas is led by the power generation and industrial sectors. Additions to gas-fired power generation capacity in the United States have outstripped electricity demand growth over the last 5 years and the EIA expects that the utilization of this capacity will rise as electricity demand continues to grow.

Coal consumption in the United States is projected to increase by 1.5 percent per year. About 90 percent of the coal is currently used for electricity generation. The increase in coal use reflects higher utilization rates at existing plants as well as the addition of new plants.

Nuclear energy use is also expected to grow in the United States but at a very slow rate. The EIA's AEO2005 projection includes life extension and capacity upgrades at existing facilities but the EIA does not expect new plants to function economically before 2025. It may be noted that the Energy Policy Act of 2005, which was passed in August 2005, provides production tax credits for the first 6 gigawatts (GW) of new nuclear capacity; and the EIA expects that this law will be sufficient to stimulate construction of 6 GW of new capacity starting in 2013. The total electricity demand is projected to increase at an average rate of 1.8 percent per year. Rapid growth in electricity use is expected for computers, office equipment and a variety of electrical appliances in the end-use sectors. About 60 percent of the demand for renewable energy in 2025 is for grid-related electricity generation. The competitiveness of renewable fuels in both the transportation and electric generation sectors is significantly affected by the prices for oil and natural gas, as well as mandates and subsidies at the state and federal level.

US Petroleum Production, Consumption and Gulf Imports

The EIA tracks oil prices in terms of the annual average US imported refiner acquisition cost (IRAC) of crude oil. In the IEO2005 reference case, IRAC prices, measured in real 2003 dollars, are projected to decline to US$25 per barrel in 2010 as new supplies enter the market. Prices then rise slowly to over US$30 per barrel in 2025. Prices for light sweet crude, comparable to West Texas Intermediate (WTI) oil, the price of which is reported in the media daily, remain about US$6 per barrel higher throughout the forecast period. However, the recent experience with higher and more volatile oil prices has heightened the uncertainty about the long-term path of crude oil prices. To reflect the uncertainty in world oil markets, the AEO 2005 includes several alternative world oil price cases with higher oil prices. The EIA High B case assumes that oil prices fall to US$37 per barrel in the year 2010 (as expressed in 2003 dollars), rising to US$48 per barrel in 2025. Again, historical price differences for light sweet crude relative to the IRAC have varied substantially and have generally been between US$1–3 higher in each year. Recent experience has seen price differences rising to as much as US$6 per barrel above the IRAC price.

In the AEO2005 reference case, total US domestic petroleum supply, which includes lease condensate and natural gas plant liquids, increases from 9.1 million barrels per day in 2003 to a peak of 9.8 million barrels per day in 2009 as a result of increased offshore production. Beginning in 2010, it begins to decline, falling to 8.8 million barrels per day in 2025.

Petroleum demand is projected to grow from 20.0 to 27.9 million barrels per day by 2025, led by demand growth in transportation fuels. Due to rising demand, the share of petroleum demand met by net imports in AEO2005 is projected to increase from 56 percent in 2003 to 68 percent by 2025. In volume terms, total US petroleum imports are projected to increase from 12.3 million barrels per day in 2003 to 20.2 million in 2025. The Gulf region's share of total gross petroleum US imports, 20.4 percent in 2003, is expected to increase to almost 30 percent in 2025; and the OPEC share of total gross imports is expected to rise from 42.1 percent in 2003 to above 60 percent in 2025. With the higher oil prices in the AEO2005 High B case, projected growth in demand is reduced, while domestic production increases. This reduces the projected import share of total US petroleum demand compared to the reference case. By 2025, under this scenario, the import share is close to its current level (Figure 7.4).

Figure 7.4

US Petroleum Production, Consumption and Net Imports (1970–2025)

(million barrels per day)

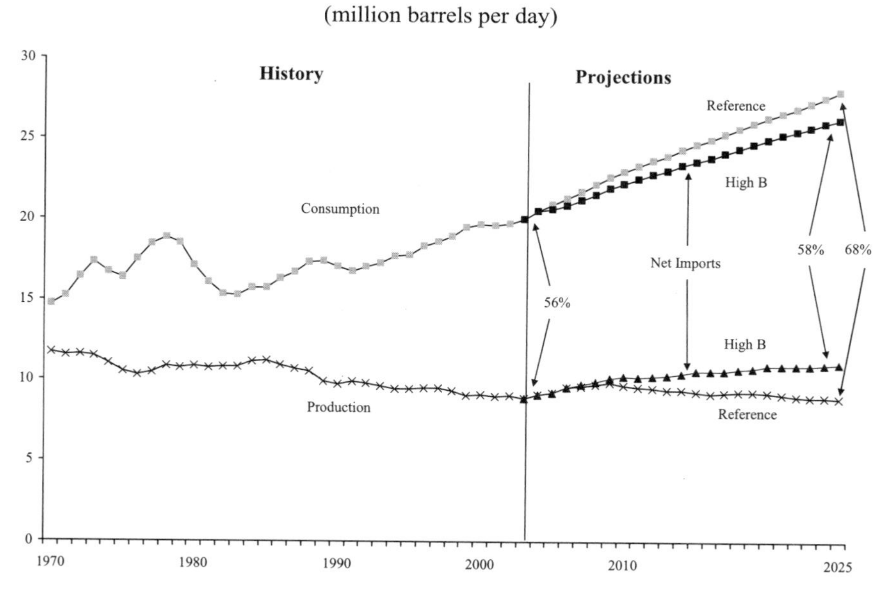

Source: *Annual Energy Outlook,* 2005.

US Petroleum Demand by Sector

Looking at domestic demand for petroleum, the transportation sector accounted for two-thirds of total petroleum demand in 2003. This share rises to 71 percent in 2025. Efficiency improvements in light-duty vehicles, trucks and aircraft are insufficient to offset growth in travel. Oil demand in the transportation sector grows from 13.4 to 19.8 million barrels per day in 2025. The industrial sector currently accounts for about one-fourth of total petroleum demand. Although its share is expected to decline, absolute consumption is expected to grow by slightly more than 1 million barrels per day over the next 20 years, from 4.9 to 6.1 million barrels per day. Demand for oil in the residential and commercial sectors grows only slightly in the forecast. In the residential sector, distillate fuel for heating is increasingly displaced by electricity, natural gas and propane. Only 3 percent of electricity generation uses oil.

In the reference case, OPEC is expected to account for less than half of the total projected gross US crude oil and petroleum product imports through 2013 and then rise to over 60 percent in 2025. Not surprisingly, almost 90 percent of the increase in imports is expected to come from OPEC member states, given their large resources and low development costs. About half of the increase in OPEC imports is projected to come from the Gulf region (the Arabian Gulf states plus Iraq and Iran) and most of the remainder from Venezuela. Crude oil imports from non-OPEC Africa are projected to increase, as Angola, Congo and Gabon add to production. Significant imports of petroleum from Canada and Mexico are expected to continue, and West Coast refiners are expected to import crude oil from the Far East to replace declining production of Alaskan crude oil. Non-conventional oil production (including production from oil sands, ultra-heavy oils, gas-to-liquids technologies, coal-to-liquids technologies, bio-fuel technologies, and shale oil) increases from 1.5 million barrels per day in 2002 to 5.7 million barrels per day in 2025 in the AEO2005 Reference case. The High B case projects that world unconventional oil would roughly double the reference case level as alternative technologies are projected to become economic and trigger substantial investments later in the projection period.

In addition to the United States, the emerging economies are likely to become increasingly dependent upon oil imports from Gulf OPEC producers as their demand for oil increases. In 2002, emerging economies imported 6.6 million barrels from the Gulf OPEC nations, which amounted to 35 percent of their total oil imports. By 2025, imports by emerging economies from Gulf OPEC producers grow to 20.0 million barrels per day or 52 percent of their total imports.

OPEC Gulf Production Capacity Additions

More than 42 million barrels per day of oil production capacity relative to the 2002 levels are expected to be added worldwide by 2025. Five of the top 12 countries projected to increase oil productive capacity are OPEC member countries: Saudi Arabia, Iraq, Kuwait, UAE and Venezuela. OPEC oil production capacity is expected to increase from 30.6 million barrels per day in 2002 to 56 million barrels per day in 2025. The largest additions to oil production capacity among the Gulf producers is projected to occur in Saudi Arabia, where productive capacity grows by 7.1 million barrels per day over the forecast period. Substantial increases in production capacity are also expected in other Gulf countries, notably Iraq (4.6 million barrels per day), the United Arab Emirates (2.5 million barrels per day), and Kuwait (3.1 million barrels per day). Among the other OPEC countries, oil production capacity increases by 6.8 million barrels per day, from 9.9 million barrels per day to 16.7 million barrels per day. Non-OPEC suppliers are expected to increase productive capacity by 16.8 million barrels per day between 2002 and 2025.

OPEC Reserve Estimates

Oil resources are adequate to meet growing oil demand to 2025. OPEC accounts for 69 percent of the world's proven oil reserves, and six of the fifteen countries with the largest proven reserves are OPEC members and collectively account for 61 percent of the world's oil reserves. According to the *Oil & Gas Journal* of December 20, 2004, OPEC accounted for 885 billion barrels of the 1,278 billion barrels of the world's proven oil reserves[3] (See Figure 7.5).

Figure 7.5

World Oil Reserves by Country (as of January 1, 2005)

(in billion barrels)

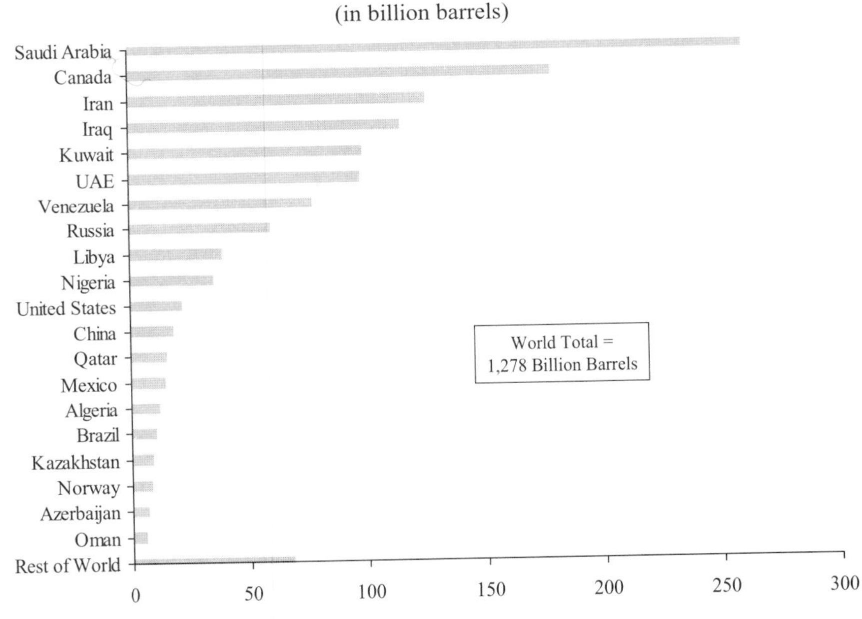

Source: "Worldwide Look at Reserves and Production," *Oil and Gas Journal*, vol.102, no. 47 (December 20, 2004).

Gulf OPEC Producers: Petroleum Cost Advantage

Production costs in Gulf OPEC nations are less than US$3 per barrel, and the capital investment required to increase production capacity by 1 barrel per day is less than US$5,940. Assuming a low price trajectory of US$21 per barrel by 2025, total development and operating costs over the entire projection period (2003–2025) would be about 24 percent of gross revenues. Thus Gulf OPEC producers can expand capacity at a cost that is a relatively small percentage of projected gross revenues.

EIA's Downward Revision of Gulf Oil Capacity Forecast

The EIA's most recent outlook for oil capacity through 2025 has been revised downward mainly in the Gulf region and precisely in Saudi Arabia. The EIA 2005 forecast of three production cases for the world is projecting oil capacity to increase to 115.5, 122.2 and 135.2 million barrels per day depending on oil prices, compared to the EIA IEO2004 forecast which projected oil capacity to increase to 126.1, 117.3 and 137 million barrels per day. In the IEO2005, the Gulf is anticipated to provide 27.8, 39.3 and 50 million barrels per day, increasing its share in the EIA's middle case to 32 percent. In the IEO2004, the Gulf was anticipated to provide 45, 32.9, and 56.8 million barrels per day. The EIA forecast IEO2005 projects that Saudi Arabia will produce 16.3 million barrels per day by 2025 or 6.2 million barrels per day less than the previous forecasts, mainly due to a sustained crude oil price increase. The previous EIA forecast IEO2004 projected a similar volume (16 million barrels per day) under a high oil price scenario. In this downward revision, EIA has mainly adopted the IEO2004 High Oil Price Case volume of the Gulf and Saudi Arabia oil production capacity as the reference case for the IEO2005 in view of the current oil price increase.

The current price increase began in 2004, when crude oil prices almost doubled from 2003 levels, rising from about US$30 per barrel at the end of 2003 to peak at US$56.37 on October 26, 2004. After falling back briefly, prices then continued to rise in 2005. This is a significant change from what we experienced during much of the 1980s and 1990s. For most of the time since the early 1980s, we have lived in a market in which spare capacity existed for crude oil production, refining and delivery systems. Crude oil suppliers outside the Organization of Petroleum Exporting Countries (OPEC) produce at maximum rates (with no surplus production capacity) for economic reasons. Thus, the world's surplus crude oil production capacity resides in OPEC (mainly in Saudi Arabia).

The large growth in non-OPEC capacity and production in areas like the North Sea and Alaskan North Slope, along with softening demand from high prices, led to major cuts in OPEC production in the 1980s, creating large capacity surpluses. As demand grew through the 1990s, OPEC production increased, but new productive capacity was not added. Short-term imbalances between supply and demand occurred and the world experienced some price swings, but those imbalances did not last long, as capacity generally existed to remedy the situation within a year. During most of the 1990s, the West Texas Intermediate (WTI) crude oil price averaged close to US$20 per barrel, but plunged to almost US$10 per barrel in late 1998 as a result of demand growth slowing down due to the Asian financial crisis just as extra supply from Iraq was entering the market for the first time since the Gulf War. OPEC producers reacted by reducing production and crude oil prices not only recovered but increased to about US$30 per barrel, as demand grew in the face of OPEC production discipline.

Beginning in 2004, world oil demand growth accelerated significantly. For the 10 years prior, from 1994 to 2004, world oil demand growth had averaged 1.2 million barrels per day. However, in 2004, world demand jumped by 2.6 million barrels per day, led by an unprecedented increase in demand from China of about 1 million barrels per day, compared to that country's increase of 0.4 million barrels from 2002 to 2003. This unusually rapid growth in demand, along with growth in the United States and the rest of the world, quickly used up much of OPEC's available surplus crude oil production capacity. As the world balance between supply and demand tightened considerably, ongoing supply uncertainties associated with Russia, Iraq and Nigeria heightened market concerns over the availability of crude oil and prices rose. In 2005, Iran, Ecuador and Venezuela added new uncertainties.

Global oil demand is expected to grow more slowly during the years 2005 and 2006, increasing by about 1.7 to 1.8 million barrels per day. China's demand is projected to increase by 0.5 million barrels per day and US demand by 0.4 million barrels per day during 2006. Together, these two countries are projected to account for about 50 percent of the world's growth in petroleum demand next year. Crude oil production capacity increases are expected to keep pace with these demand increases. Production increases from OPEC members are projected to represent almost one-third of the world production growth next year, and the former Soviet Union is expected to provide an additional 40 percent of the increase. Other areas such as the United States and other non-OPEC countries will provide additional production volumes. However, the EIA is not projecting much increase in the surplus capacity cushion any time soon. Spare capacity was projected to remain at or below 1.2 million barrels per day in 2005, mostly heavy crude that is not useful to non-complex refineries for turning into light products (See Figure 7.6).

Figure 7.6
OPEC Spare Capacity is Extremely Tight

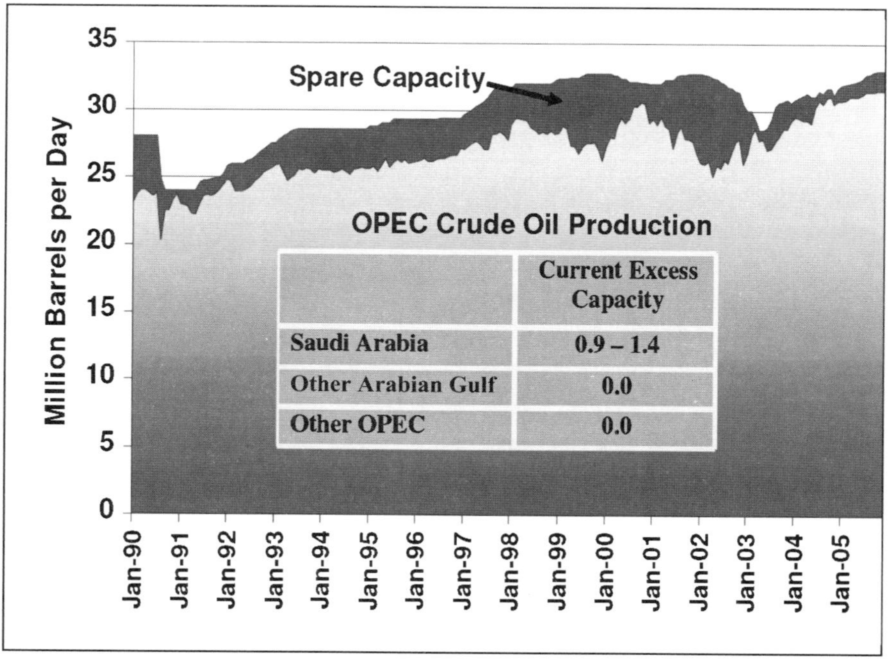

Spare Capacity

OPEC Crude Oil Production

	Current Excess Capacity
Saudi Arabia	0.9 – 1.4
Other Arabian Gulf	0.0
Other OPEC	0.0

Million Barrels per Day: 0, 5, 10, 15, 20, 25, 30, 35

Jan-90, Jan-91, Jan-92, Jan-93, Jan-94, Jan-95, Jan-96, Jan-97, Jan-98, Jan-99, Jan-00, Jan-01, Jan-02, Jan-03, Jan-04, Jan-05

Source: Energy Information Administration (EIA) estimates.

The world is facing tight crude oil markets for a number of years. The EIA's Short-Term Energy Outlook projects that WTI crude oil prices will remain above US$55 through the year 2006. Even if demand softens or capacity develops faster than anticipated, statements from OPEC members indicate an intention to keep prices from falling below US$50 per barrel. While this price is high in relation to recent years, the price of crude oil import in 2004, adjusted for inflation, was US$33.26, still below the levels seen in the early 1980s, when it reached US$62.71 in 1980.

This tight balance results in different behavior and price implications than those exhibited by the short-term market imbalances seen over the past 20 years. Instead of high prices being accompanied by low inventories and expectations for prices to be falling quickly in the future, today, in both crude oil and product markets, we see high prices with high inventories.

US Natural Gas Production, Consumption and Net Imports

US domestic natural gas production is expected to rise more slowly than consumption over the forecast period, rising from 19.1 trillion cubic feet in 2003 to 21.8 trillion cubic feet in 2025. The difference between consumption and production is made up by increasing reliance on natural gas imports. In AEO2005, net imports grow from 3.2 trillion cubic feet in 2003 to 8.7 trillion cubic feet in 2025. There is a major change in the source of US natural gas imports over the forecast period. Pipeline imports from Canada are expected to decline both in absolute terms and as a share of total imports. However, imports of liquefied natural gas (LNG) increase rapidly. LNG imports, which supplied less than 2 percent of the US market in 2003, are projected to supply over 20 percent of total US natural gas demand in 2025 (See Figure 7.7).

According to the AEO2005 highest price case High B, by 2025, domestic natural gas production is expected to increase more slowly than consumption over the forecast, rising from 19.0 trillion cubic feet in 2003 to 23.5 trillion cubic feet in 2025 versus 21.8 trillion cubic feet in the reference case. The difference between consumption and production is made up by increasing imports, particularly LNG, with a 2.7-trillion cubic feet versus 6 trillion cubic feet in reference case net increase expected over 2003 levels. By 2025, the EIA expects expansion at three of the five existing LNG import terminals and construction of several new terminals.

Figure 7.7
US Natural Gas Production, Consumption and Net Imports (1970–2025)
(in trillion cubic feet)

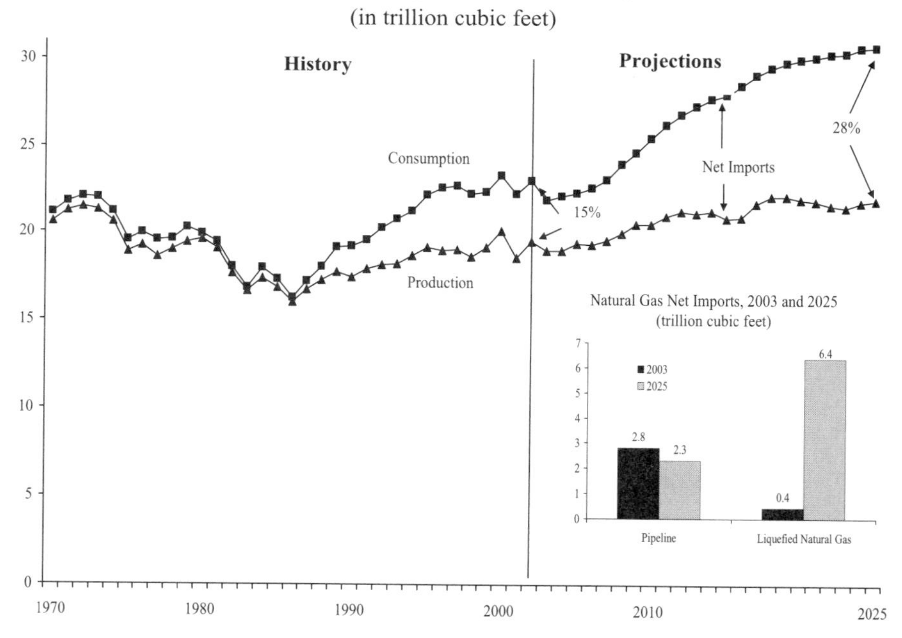

Source: *Annual Energy Outlook*, 2005.

US Natural Gas Demand By Sector

Total natural gas demand is expected to increase from 21.9 to 30.7 trillion cubic feet between 2003 and 2025. Growth in demand for electricity generation and industrial applications accounts for about 75 percent of the projected growth in total natural gas demand. The projections for electric power sector demand are sensitive to projected prices. If prices are higher than assumed in the AEO2005 reference case, electric power demand for gas will likely be lower.

The strongest growth in natural gas consumption is in the electric power sector, where consumption is projected to almost double, from 4.9 trillion cubic feet in 2003 to 9.4 trillion cubic feet in 2025. Demand by electricity generators is expected to account for over 30 percent of total end-use natural gas consumption in 2025, as compared with about 23 percent in 2003. Electric power gas consumption growth results from both the construction of new gas-fired generation plants and from a higher capacity utilization of current gas-fired plants. Most new electricity generation capacity is expected to be fueled by natural gas, because gas-fired generators are projected to have advantages over coal-fired generators that include lower capital costs, higher fuel efficiency, shorter construction lead times and lower emissions. Towards the end of the forecast period, however, when natural gas prices rise substantially, coal-fired power plants are expected to be competitive for new capacity additions, and gas begins to lose market share to coal.

Industrial consumption (including lease and plant fuel) remains the largest consuming sector and is projected to increase from 8.1 to 10.3 trillion cubic feet between 2003 and 2025. Those industrial sectors projected to experience the greatest gas consumption growth include metal-based durables, petroleum refining, bulk chemicals and food. In the residential and commercial sectors, natural gas consumption is projected to increase by slightly under (0.7 percent) and slightly over (1.2 percent) 1 percent per year respectively, from 2003 to 2025.

According to the AEO2005 High B scenario, US natural gas consumption is expected to increase at an average annual rate of 1.1 percent between 2003 and 2025. Natural gas consumption by electric generators is expected to increase by about 45 percent over the forecast, from 5.1 trillion cubic feet in 2003 to 7.4 trillion cubic feet in 2025, which is an average annual growth rate of 1.7 percent. Demand by electricity generators is expected to account for 26 percent of total natural gas consumption in 2025. Most new electricity generation capacity is expected to be fueled by natural gas, which is expected to be favored over coal. The industrial sector is the largest natural gas-consuming sector, with significant amounts used in the bulk chemical and refining sectors. Industrial consumption, excluding lease and plant fuel, is expected to increase by 2 trillion cubic feet over the forecast period, driven primarily by macro-economic growth. Combined consumption in the residential and commercial sectors is projected to increase by 1.7 trillion cubic feet from 2003 to 2025, driven by increasing population and economic growth. Natural gas remains the overwhelming choice for home heating throughout the forecast period.

Supply of Natural Gas to the US Market

Total natural gas supply is projected to increase at an average annual rate of 1.4 percent per year between 2003 and 2025, from 22.4 to 30.6 trillion cubic feet. Existing sources of US natural gas production are projected to decline from today's output levels over the next two decades. Future US natural gas supplies will depend on natural gas from Alaska, unconventional domestic production that includes tight sands gas (produced from low-permeability reservoirs) coal-bed methane and LNG. Natural gas pipeline from Alaska is projected to enter service by 2016.

Under the AEO2005 High B oil price case, increasing natural gas production is supported by rising wellhead natural gas prices, and improvements in production technologies, particularly for unconventional natural gas. The national average wellhead price is projected to reach US$5.32 per thousand cubic feet (mcf) in 2003 dollars by 2025. Increased US natural gas production through 2025 comes primarily from unconventional sources and from Alaska. Unconventional natural gas production increases by 2.3 trillion cubic feet over the forecast period, largely because of expanded tight sands gas production in the Rocky Mountain region. Annual production from unconventional sources is expected to account for 38 percent of production in 2025, more than any other source, compared to 35 percent today. Alaska accounts for most of the growth in domestic natural gas production, growing by 2.6 trillion cubic feet over the forecast period. Lower 48 onshore non-associated conventional natural gas production declines by about 650bcf through 2025, as resource depletion causes exploration and development costs to increase. The natural gas share of electricity generation is projected to increase from 16 percent in 2003 to 24 percent in 2025. The natural gas share peaks at 26 percent in 2018 and then declines to 24 percent in 2025. Renewable technologies grow slowly because of the relatively low costs of fossil-fired generation and because competitive electricity markets favor less capital-intensive technologies. Where enacted, state renewable portfolio standards, which specify a minimum share of generation or sales from renewable sources, are included in the forecast. Nearly all of the existing nuclear, renewable, combined cycle and coal steam capacity now in use is projected to remain in use through 2025. The only significant projected retirements are among steam plants burning oil and natural gas. Cumulative retirements of existing units total 43 gigawatts. Between 2004 and 2025, over 281 gigawatts of new generating capacity is added, including end-use sector combined heat and power capacity additions of 18 gigawatts. Much of the new capacity is added in the Southern tier of states due, in part, to population migration to the South. The majority of projected new generating capacity additions are expected to be natural gas-fired units, particularly in the early years of the forecast. Between 2004 and 2025, 178 gigawatts of natural gas-fired capacity is added. After the year 2015, coal is much more competitive with respect to new generating capacity. Between 2004 and 2025, around 87 gigawatts of coal capacity is added.

Price of Natural Gas in the Annual Energy Outlook 2005 Reference Case

The average wellhead prices for natural gas are projected to increase through 2005 and then decline to US$3.64 per thousand cubic feet in 2010 as the initial availability of new import sources and production from increased drilling expands the available supply. After 2010, wellhead prices are projected to increase gradually, reaching US$4.79 in 2025. In nominal dollars, the 2025 price is the equivalent of US$8.23 per thousand cubic feet. The increase is in response to higher exploration and development costs associated with smaller and deeper gas deposits in the remaining domestic gas resource base. Growth in unconventional sources, Alaska production, and LNG imports are not expected to increase enough to offset the impacts of resource depletion and increased demand. Prices are projected to increase in an uneven fashion as new, large volume supply projects (such as an Alaska pipeline) temporarily depress prices when initially brought online.

Price of Natural Gas in the Annual Energy Outlook 2005 High B case

In the Annual Energy Outlook 2005 High B case, average natural gas wellhead prices are projected to decline from current high levels to US$3.74 per mcf in 2010, in 2003 dollars, due to expanded imports and production of unconventional and offshore natural gas. After 2010, wellhead prices are projected to increase gradually, reaching US$5.32 per mcf in 2025 in 2003 dollars. Ultimately, the growth in unconventional sources, Alaskan production, and LNG imports are not expected to increase enough to offset the effects of resource depletion and increased demand. End-use natural gas prices are expected to reflect the trend of increasing wellhead prices. A portion of the increase in wellhead prices is expected to be offset by a projected decline in average transmission and distribution margins as a larger proportion of the natural gas delivery infrastructure becomes fully depreciated. Residential consumers, who generally pay the highest costs for distribution infrastructure, will see the largest offset.

US Import Terminals for LNG

There are currently five LNG import terminals in the continental United States, at Elba Island, Georgia; Cove Point, Maryland; Everett, Massachusetts;

Lake Charles, Louisiana; and the offshore Gulf of Mexico. These terminals currently have an estimated combined peak capacity of about 1.8 trillion cubic feet per year and an estimated annual base-load capacity of 1.3 trillion cubic feet. The Gulf Gateway Energy Bridge in the offshore Gulf of Mexico came online in March 2005, and is the first new receiving terminal to come online in the United States for over 20 years. With additional expansions that have been approved by the Federal Energy Regulatory Commission (FERC) for both Lake Charles and Elba Island, and the completion of 2 terminals that have broken ground at Freeport, Texas and Sabine Pass, Louisiana, the peak annual capacity could reach over 3.5 trillion cubic feet. The potential for additional expansion depends on a variety of site-specific factors, such as the availability of additional land and harbor constraints on tanker traffic and on the number of tankers that can be berthed simultaneously. Expansion of existing facilities is less costly than the construction of new facilities, and it is assumed that additional expansion, within limits, can occur at existing facilities. This does not include Everett, where there is little room for physical expansion due to its location (See Figure 7.8).

Figure 7.8
Current US LNG Import Terminals

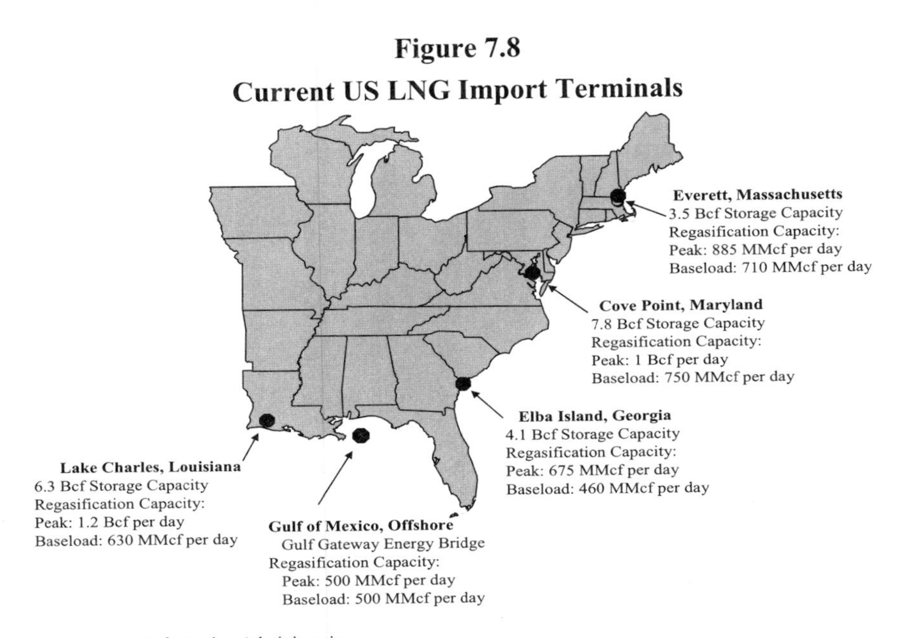

Source: Energy Information Administration.

Imports are expected to be priced competitively with domestic sources of natural gas, and net imports of natural gas are expected to make up the difference between US production and consumption. LNG is expected to account for most of the projected increase in net imports. By the end of the forecast, sufficient new LNG terminal capacity comes online to allow net LNG imports to increase from 440 billion cubic feet (bcf) in 2003 to 6.4 trillion cubic feet in 2025. By 2015, net LNG imports are expected to equal 15 percent of total US gas consumption, compared to 2 percent in 2003. Net LNG imports are expected to rise from 13 percent of net imports in 2003 to 62 percent in 2025.

Until a few years ago, Algeria was the largest supplier of LNG to the United States. Since 2000 it has been far surpassed by Trinidad and Tobago, which was the source for 75 percent of the nation's gross LNG imports in 2003. In addition to Trinidad and Tobago and Algeria, the United States also received LNG cargoes from Malaysia, Nigeria, Oman, and Qatar in 2003. According to the September 2005 EIA Short-Term Energy Outlook, gross LNG imports for 2005 were expected to be 860 bcf and LNG imports for 2006 were expected to be 1,210 bcf. With considerable expansion of liquefaction capacity currently underway and also planned for the future, diversification in supply sources for the United States is expected to grow. Qatar, Egypt, Norway, Equatorial Guinea and Russia have liquefaction capacity under construction, and at least seven other countries are planning to become LNG exporters. These countries include Bolivia, Peru, Angola, Yemen and Iran.

Net imports of natural gas from Canada are projected to be 3.0 trillion cubic feet in 2005, declining gradually to 2.5 trillion cubic feet in 2009. A MacKenzie Delta natural gas pipeline is projected to begin transporting gas in 2010, and imports are expected to rise subsequently to 3.0 trillion cubic feet in 2015. After 2015, net gas imports from Canada are projected to decline again, falling to 2.5 trillion cubic feet in 2025. Conventional production in the Western Sedimentary Basin is projected to decline throughout the projection period but unconventional gas production in

Western Canada, conventional production in the MacKenzie Delta and Eastern Canada, and liquefied natural gas imports are expected to more than offset the production decline in the Western Sedimentary Basin. The reason for the decline in Canadian imports towards the end of the forecast period is that Canadian gas consumption is expected to increase at a faster rate than Canadian gas production (See Figure 7.9).

Figure 7.9

Net US Imports of Natural Gas (1970–2025)

(in trillion cubic feet)

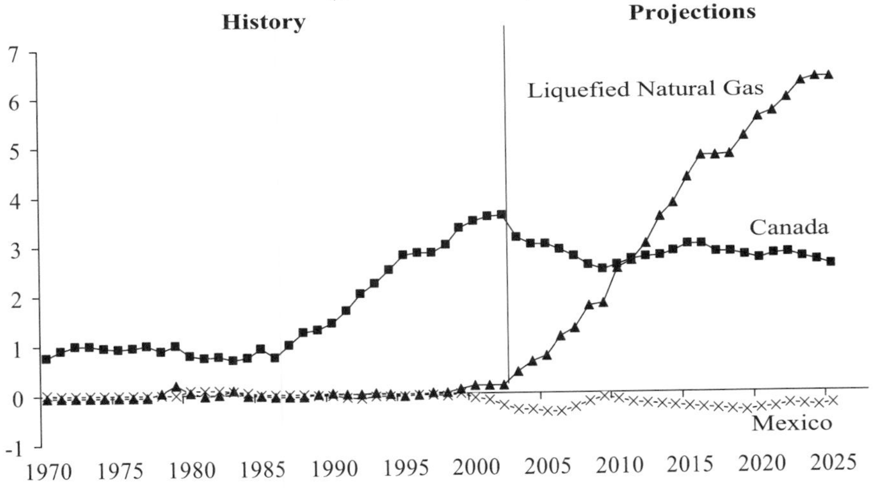

Source: *Annual Energy Outlook*, 2005.

Although Mexico has considerable natural gas resources, the United States historically has been a net exporter of gas to Mexico. Net exports of US natural gas to Mexico are projected to grow until 2006, and subsequently decline after 2006 as LNG terminals in Baja California come online to serve both the Mexican and US markets.

LNG Imports to the US under High B Oil Price Scenario

Under the High B Scenario, net imports of natural gas, primarily in LNG and sourced from Canada, are projected to increase less than in the reference case, from 3.2 trillion cubic feet in 2003 to 5.7 trillion cubic feet in 2025 versus 8.7 trillion cubic feet in 2025. Imports contributed 15 percent to total natural gas supply in 2003, compared to an expected 20 percent in 2025 (See Figure 7.10). LNG is expected to supply all of the increase in US imports. LNG imports are expected to reach 3.6 trillion cubic feet in 2019 and then decline to 3.1 trillion cubic feet in 2025 less than the 6.4 trillion cubic feet under Reference Case, as rising natural gas prices cause a decline in demand. In 2025, LNG is expected to equal 11 percent of total US supply. Net Canadian imports are expected to provide 10 percent of total US supply in 2025, down from 14 percent in 2003.

Figure 7.10
Natural Gas Production, Consumption and Imports (1970–2025)
High B Oil Price Case (in trillion cubic feet)

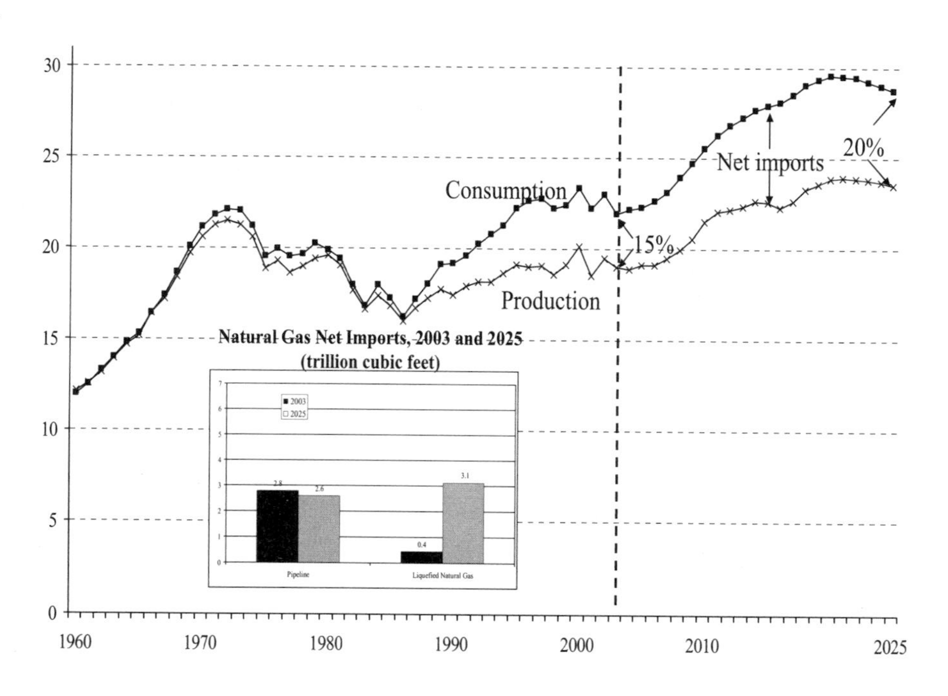

Source: *Annual Energy Outlook*: High B Oil Price Case

US Imports of LNG from the Gulf: Future Predominance of Qatar

From the signed contracts, Qatar will produce 68.9 million metric tons per annum of LNG by 2010, positioning this Gulf state as the world's leading supplier. About 23.4 million metric tons per annum is destined for the US market through ExxonMobil and ConocoPhillips. The US markets will import 2.5 trillion cubic feet of LNG or the equivalent of 52 million metric tons per annum of LNG in 2010. Over 40 percent of US LNG needs could come from the Gulf region (under the reference case and High B price case scenario) because Qatar can penetrate any market of its choice and cover its cost of gas production, which is extremely low (almost zero). Production is backed by the condensates associated with gas from the 900 trillion cubic feet North Field, projected at 500,000 bpd by 2010. Currently the Qataris charge themselves 50–75 cents per million Btu (MMBtu). Qatar has also benefited enormously from the economies of scale and technology cost reduction achieved in their liquefaction capacity.

By 2010 Qatar is poised to become one of the world's leading producers of liquefied natural gas (LNG). The country has been very successful in finding new markets. In 2002, Qatar earned around US$3.7 billion from exporting 15 million metric tons of LNG.[4] At the LNG Ministerial Summit sponsored by the US Department of Energy and held in Washington, DC in December 2003, Qatar's Second Deputy Prime Minister and Minister of Energy and Industry, Abdullah Bin Hamad Al Attiyah, announced that his country would invest around US$25 billion in LNG projects by 2010, quadrupling its export capacity.[5]

Qatar is a relatively new supplier, sending its first LNG shipment to Japan in 1997. Its focus is still the Asian market, the proximity of which has been strategic in ensuring profitability. As technology has reduced the cost of liquefaction and shipping by almost a third in the last few years, Qatar has become the new focus of attention as it negotiates projects that will expand its market share in Asia and allow it to enter the US markets in a big way.

With proven reserves of over 900 trillion cubic feet, Qatar's natural gas resources rank third in size behind those of Russia and Iran.[6] Most of the country's reserves are located in the North Field, to date the largest known non-associated gas field in the world. Qatar began developing the North Field gas reserves in 1984, mostly producing condensates. Condensate is a light hydrocarbon liquid suspended in natural gas reservoirs and can be recovered by condensation of hydrocarbon vapors. After it is separated from gas, it remains liquid without either pressurized or refrigerated containment. In addition, the Dukhan field and smaller associated gas reserves in the Id al Shargi, Maydan Mahzam, Bul Hanine, and Al-Rayyan oil fields are estimated to contain 10 trillion cubic feet of gas.

For the last few years the Qataris have opted to diversify their gas portfolio by investing in regional gas pipeline projects, gas-to-liquid technology, and the expansion of their liquefaction capacity. The most ambitious regional pipeline project to date is the US$4 billion Dolphin Gas Project that will pipe gas over 260 miles from Qatar to the UAE and Oman, delivering an estimated 2 billion cubic feet per day by 2006.[7] Though importing countries have their own reserves, and export LNG themselves, they find it less costly to import Qatari gas than to develop and treat their own non-associated supplies. Kuwait and Bahrain, two other Gulf states, have also approached Qatar with a view to following suit.

Qatar has also invested in gas-to-liquid technology (GTL). This approach, developed at great cost, converts natural gas into high-grade gasoline and distillates. Qatar has already drawn up plans to produce 800,000 barrels per day from gas.[8]

In 2003, Qatar signed two agreements, the first, with ExxonMobil, to provide the United Kingdom with 15 million metric tons per annum by 2006/7, the second, with ConocoPhilips, taking up 9.2 million metric tons per annum, some 7.5 million metric tons per annum of this by 2008/9, destined for the United States. ExxonMobil too is working on an additional 15 million metric tons per annum for the United States by

2010.[9] In all these projects, Qatar intends partnering international companies across the entire spectrum, from production to liquefaction, transportation, regasification and distribution.

Qatar LNG is expected to occupy a leading position in the United States market over the next two decades. A recent study by Cambridge Energy Research Associates (CERA) concluded that the presence of a large volume of LNG imports in the United States market is indispensable. For the next five years, with a price range of gas expected to be higher than US$4 per million btu, Qatar will be in a position to recover its costs in the United States market.[10]

The State of Qatar has the lowest exploration and development costs for gas of any region in the world, with capital costs estimated at less than US$0.20 per million Btu.[11] Even though most of Qatar's gas is offshore, the transmission pipelines to connect the gas fields to the LNG liquefaction plants are relatively short, comprising only a small share of the overall cost. An added bonus is that most of the proposed liquefaction projects are in Ras Laffan Industrial City, where they take advantage of existing infrastructure and large areas of unused land for future expansion. Technological advances are such that the capacity of the new trains might reach 7 million metric tons per annum (prior to this the limit was 2 to 3 million metric tons). Once economies of scale are factored in, the competitiveness of Qatar's LNG is expected to continue for years to come. In order to finance its current projects, Qatar maintains and enjoys a strong credit rating, despite regional unrest.

Although LNG imports comprise less than 2 percent of the US market, this amount will increase substantially over the next two decades. At present, the United States has five terminals that receive LNG. By 2025, the projected increase is estimated at 14 billion cubic feet per day, necessitating at least 9 more terminals. In fact the major challenge regarding the future of LNG in the United States is not the availability of terminals (a need that is slowly being met), but rather the reliability of supply. Equally important, is the matter of transparent and sustainable rules governing the gas business per se.[12]

A supplier such as Qatar, which is willing to invest billions in enlarging liquefaction capacity to supply the US markets, must give thought to the US-Algerian experience of the late 1970s. In this instance, Algeria had become the first country to export LNG to the States. At the time four terminals were built at the eastern seaboard, their purpose being to regasify Algerian LNG. Pricing was cost-based, and regulated accordingly, with the result that prices were held below market-clearing levels, stimulating demand but inhibiting domestic investment in new production. The Canadians, for example, took advantage by setting a "uniform border price" of US$4.64 for exports to the United States, and did so at a time (1976) when the US wellhead price was US$0.58. Although in theory LNG imports benefited from cross-subsidizing[13] when US regulations changed, imports were opened up to competition. Not surprisingly, the bottom dropped out of the market as American drilling surged again. The prices negotiated by US importers – long-term contracts for Algerian LNG – were no longer viable, and firms that had handled regasification found themselves bankrupt and closures began.[14] The terminals at Cove Point, Maryland and Elba Island, Georgia, built in 1978, stopped business in 1980. The terminal at Lake Charles, Louisiana opened in 1980, closed down within three years, and although the one in Everett, Massachusetts operated from 1971 until 1985, eventually, it too shut down for a short period of time.

Some commentators blame this unfortunate business on the rigidity of Algerian contracts, a claim with some basis in fact. In the wake of OPEC's success in raising oil prices in the seventies, Algeria began demanding a price for its LNG that was comparable to the price of oil. This meant prices at the tailgate of the US terminals would be around US$7.00 per thousand cubic feet (Mcf),[15] with the result that Algerian gas could not compete in a market that was well supplied with domestic gas. LNG deliveries to the United States dwindled.

These days, the situation in North America is markedly different. Growth in gas production has fallen since 2001. Despite drilling efforts, domestic US producers are not expected to arrest this decline, nor will they be in a position to develop gas within the next five years for less than US$4 per million British Thermal Units (MMBTU),[16] which is the LNG price that allows Qatar to recover costs in the US market.

A major concern to a supplier such as Qatar is the unpredictability of the United States moves to deregulate the gas and electric power industries. LNG suppliers see deregulation negatively because it is likely to result in changes to the business environment, such as the insistence on shorter contracts, the removal of the take-or-pay and fixed destination clauses from future contracts, and requiring third-party access to regasification facilities. These changes force suppliers to shoulder a greater portion of the risk, which might hinder the development of liquefaction facilities.

The US Federal Energy Regulatory Commission (FERC), which has jurisdiction for LNG pricing and siting, has lately eased its requirement for open access to regasification capacity. This development has encouraged potential suppliers such as Qatar and its major partners to consider investing in new terminals.

At present, many LNG investors who have been monitoring the Henry Hub index of natural gas prices are eager to capture what appears to be a high margin of profitability in supplying LNG to the US market. Although Henry Hub index prices have been higher than the cost of LNG imported to the United States for the last past five years, some observers believe that the index does not reflect market realities, and may encourage over-investment in LNG that will not be economically sustainable. A supplier as large as Qatar, willing to invest today in order to deliver LNG through 2010, might end up being the swing producer in case US demand slows down.

At this point, it is unclear whether Qatar will be willing to accept the high financial risks associated with increasing its LNG capacity to supply the North American market. Expanding LNG capacity is relatively expensive. In the oil industry, it is estimated that a 1 percent capacity surplus in worldwide oil production and maintenance (assuming an investment of US$5,940 per daily barrel) would cost about US$2.2 billion. By contrast, creating a 1 percent surplus in world-wide gas production, assuming that the entire 1 percent was brought to market as LNG plus storage would have a price tag of about US$13.8 billion (assuming an investment of $3.85 per annual Mcf in gas field development, plus liquefaction, tankers and regasification costs).[17] How prepared Qatar is to accept the high financial risks associated with supplying the US markets market remains a question.

EIA's Key Assumptions in Forecasting the US–Gulf Energy Trade

The National Energy Modeling System (NEMS) is the model used by the Energy Information Administration (EIA) to prepare the *Annual Energy Outlook 2005* (AEO2005), and the System for the Analysis of Global Energy Markets (SAGE) is the model used by the EIA to prepare the *International Energy Outlook 2005* (IEO2005) mid-term projections. These models start with some key assumptions in order to model global energy supply, demand and trade among regions such as the United States and the Gulf. The current SAGE framework uses historical data on fuel consumption from EIA's International Energy Annual 2002 to calibrate the base year for the model's Reference Energy System (RES). SAGE also aggregates individual countries into 15 fixed global regions and forecasts all energy consumption activity and consequent emissions at the regional level. The Arabian Gulf states are within the Middle East grouping that also includes Iran and Turkey. SAGE derives demand for energy end-use services (residential space heating, personal road transportation, etc.) as a function of exogenously developed macroeconomic drivers (GDP, population, etc.) The SAGE linear program then calculates the least-cost supply and technology mix that will meet the projected demand. Although SAGE does not endogenously change the projections provided exogenously for the macroeconomic drivers, most of the effect of higher energy prices is captured in the elastic demand. Furthermore, higher GDP growth projections tend to increase the introduction of newer, more efficient technologies. In addition, SAGE imposes minimum market shares for regional supply and some energy-efficient technologies (for example, newer, more efficient personal automobiles, such as gas-electric hybrids) that have capital costs higher than those of less efficient competing technologies. (Because of the "winner-take-all" nature of linear program models, the more efficient but more costly technologies would not be selected without the imposed minimum shares.)

Limitation of Modeling to Predict Future of US–Gulf Energy Trade

It is important to recognize that any modeling effort has its limitations, and the data generated from the EIA models are not intended to predict what will happen but what might happen, given known technology, technological and demographic trends and current laws and regulations. Thus these models provide a policy-neutral data for analysis. Also such energy market projections are subject to much uncertainty, including political disruptions, severe weather and technological breakthroughs. While it is certain that the trends in energy trade and markets for the US–Gulf States will be similar to what is listed in the *Annual Energy Outlook* and the *International Energy Outlook*, it is less certain whether the exact numbers suggested by these two publications will represent the final reality.

Section 4

FUTURE DEMAND FOR GULF OIL

8

Supplying Asia-Pacific Oil Demand: Role of the Gulf

Kang Wu and Jit Yang Lim

The Asia-Pacific is one of the fastest growing regions in the world both in economic terms and in terms of energy consumption.[1] However, the region is poor in resources, particularly in oil and gas, and is heavily dependent on imported energy. The oil import level alone has already made the Asia-Pacific the largest energy-importing region in the world. Over the next ten years and beyond, Asia's energy imports, especially oil imports, will continue to grow. Within the region, China is one of the most dynamic countries with growing energy demand and rising oil and gas imports. For Asia in general, and China in particular, the Gulf plays a critical role in supplying oil and gas. As their energy import dependences deepen, the Gulf will assume greater importance.

This chapter analyzes oil demand and supply in the Asia-Pacific region, examines the rapidly changing energy and oil markets in China, and assesses the potential of the Gulf in meeting the surging oil demand in Asia and China. The first section briefly reviews the structure of energy consumption and fossil energy import dependence of the Asia-Pacific region. The second section presents the outlook for petroleum product demand and oil production in the Asian continent. The rise in Asian oil imports and the future role of Gulf are assessed in the third section. The final section studies China's growing energy demand, deepening energy import dependence and the Gulf region's contribution. This is followed by some concluding remarks.

Asia-Pacific: Consumption Structure and Import Dependence

In this section, two issues will be discussed. First, the unique structure of energy use in the Asia-Pacific region and how it differs from the rest of the world. Second, the main imbalances between consumption and production for individual primary energies in the region.

Unique Characteristics of Asian Energy Use

In 2004, the primary commercial energy consumption (PCEC) in the Asia-Pacific reached 64 million barrels of oil equivalent per day (boepd), up from 23 million boepd in 1980 and 36 million boepd in 1990.[2] The region's growth in energy consumption during the past two and half decades was well above the world average. This is clear from the rising share of the Asia-Pacific in the world's total primary commercial energy consumption: 18 percent in 1980, 22 percent in 1990 and 31 percent in 2005.

The unique structural features of Asia-Pacific primary energy consumption can be viewed from several aspects. The Asia-Pacific region has a large coal share in the total PCEC. In 2004, coal accounted for 47 percent of the total PCEC (See Figure 8.1), substantially higher than the level of 27 percent for the whole world (See Figure 8.2). This is mainly due the high share of coal use in China and India. Excluding these two countries, the share of coal for the rest of the Asia-Pacific is about the same as the world average (Figure 8.3). For oil, the share in total PCEC is slightly lower in the Asia-Pacific as compared with the whole world. Again, excluding China and India, the region has a considerably bigger oil share than the world average. However, the Asia-Pacific has a low share of natural gas, which has largely been underdeveloped and underutilized. The region has the potential to use more natural gas than other fossil-based sources of energy. Furthermore, the Asia-Pacific shares of nuclear power and hydroelectricity are also lower than the global average.

Figure 8.1
Structure of Primary Commercial Energy Consumption (PCEC) in the Asia-Pacific Region (2004)

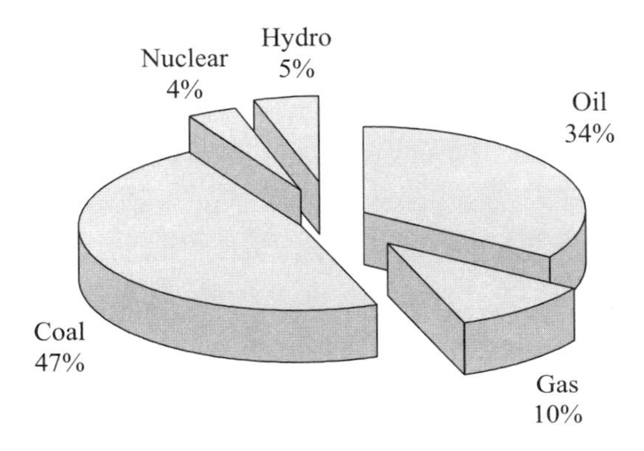

Note: Total consumption: 64 mmboepd.

Source: British Petroleum (BP), *BP Statistical Review of World Energy* (London: BP, June 2005).

Figure 8.2
Structure of Primary Commercial Energy Consumption (PCEC) for the Whole World (2004)

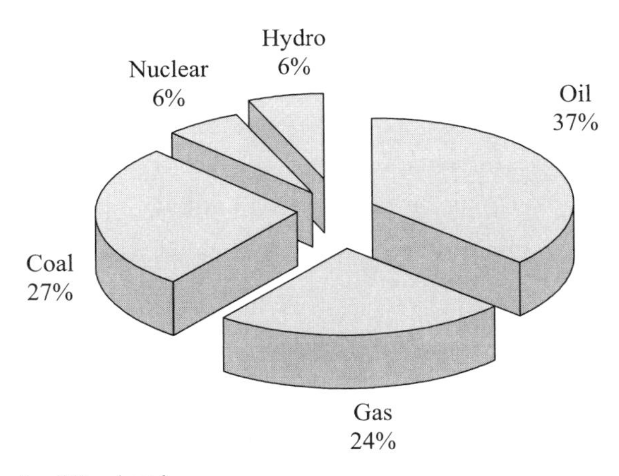

Note: Total consumption: 204mmboepd.

Source: British Petroleum (BP), *BP Statistical Review of World Energy* (London: BP, June 2005).

Figure 8.3
Structure of Primary Commercial Energy Consumption (PCEC) in the Asia-Pacific Region (excluding China and India), 2004

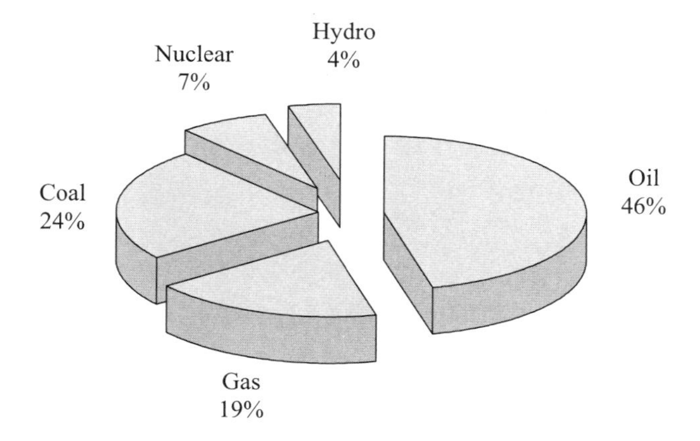

Note: Total consumption: 29 mmboepd.

Source: British Petroleum (BP), *BP Statistical Review of World Energy* (London: BP, June 2005).

Energy Import Dependence in the Asia-Pacific Region

Overall, the Asia-Pacific region has a large discrepancy in its primary fossil energy consumption and production. The shortfalls are mainly in oil although the region also needs gas imports from the Gulf. In 2004, the Asia-Pacific region accounted for 29 percent of the world oil consumption, yet its share of world oil production was only 10 percent (Figure 8.4). This gap translates into high oil import dependence. For natural gas, the Asia-Pacific region had a higher share in global consumption than production in 2004, but the gap was much narrower than oil. The region is largely balanced in coal supply, though there are large-scale intra-regional coal flows among Asian-Pacific countries.

Figure 8.4
Share of Asia-Pacific Region's Primary Fossil Energy in the World (2004)

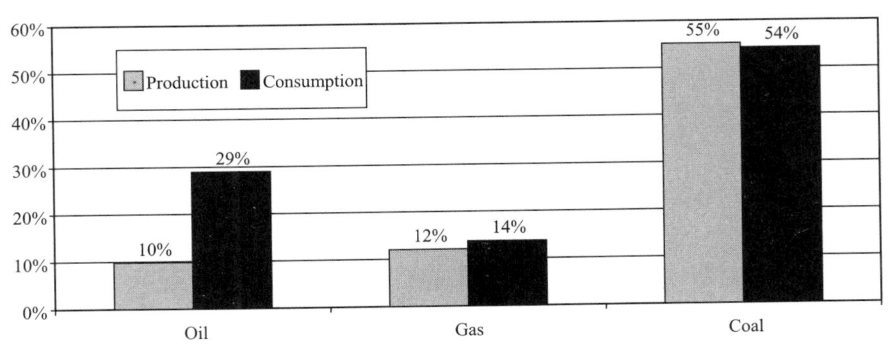

Source: British Petroleum (BP), *BP Statistical Review of World Energy* (London: BP, June, 2005).

The case of oil will be discussed further. For natural gas, the Asia-Pacific region accounts for two thirds of the global trade in liquefied natural gas (LNG). Japan is the word's largest LNG importer. Other LNG importers include South Korea, Taiwan and India, while China is on course to become the fifth LNG importer in the region. LNG imports to these countries largely come from other Asian countries, including Indonesia, Malaysia, Australia and Brunei. However, the Asian importers are getting an increasing amount of LNG from the Gulf and Russian Far East. In addition, the US state of Alaska is also a small supplier of LNG to Asia. Over the next ten to fifteen years, FACTS projections indicate that LNG imports to Asia will continue to rise at a pace faster than the growth of oil. However, since 2004, the rising oil and gas prices have slowed down the LNG demand in the region, particularly in India and China. For natural gas, the era of under $4 per million BTU (mmBTU) of ex-ship prices is over and the Asian buyers are facing the prices of US$7 per mmBTU and above. For FOB Dubai prices, the monthly prices reached US$64.14 per barrel, more than double the price of US$28.89 per barrel that prevailed in January 2004.

The Asia-Pacific region has both large coal producers (notably China, Australia, India, and Indonesia) and consumers (such as China, India, Japan, Australia and South Korea). The major regional coal exporters are Australia, Indonesia, China, Vietnam and others, while the major importers include Japan, South Korea, Taiwan, India and China. In the case of China, it is a sizable coal exporter but its imports have also been on the rise. However, on an overall basis, China is a net coal exporter but its surplus is shrinking. FACTS expects that China will eventually become a net coal importer after 2010. While the region relies mainly on itself to meet its growing coal needs, it does import some quantities of coal from Russia, the United States and other countries.

Outlook for Oil Demand and Supply in the Asia-Pacific Region

Generally speaking, the Asia-Pacific region is witnessing rapid growth in oil demand but production remains flat. However, the situation varies widely from country to country. In terms of oil demand and production, a few major players dominate the regional scene.

Outlook for Petroleum Product Demand in the Asia-Pacific Region

The Asia-Pacific region has served as the engine of growth for global oil demand. Whenever the Asia-Pacific market is weak, the global oil market weakens and whenever the Asia-Pacific market is strong, so is the global oil market. This is illustrated by Table 8.1.[3] However, Asia fared very poorly in 1998 due to the Asian Financial Crisis, growing stronger in 2000 as the continent began to stage a recovery. Rising oil prices in 2001 had affected Asian oil demand to a greater extent than the global market, due to its higher level of oil dependency and poorer consuming power.

Table 8.1
Global and Asia-Pacific Incremental Oil Demand, 1995–2004 (kbpd)

Year	World	Asia-Pacific (AP)	Share of AP
1995	1,400	939	67%
1996	1,630	864	53%
1997	1,500	791	53%
1998	420	-183	-44%
1999	1,620	957	59%
2000	680	544	80%
2001	690	70	10%
2002	600	260	43%
2003	1,770	805	45%
2004	2,900	1,106	38%

Sources: International Energy Agency, *Oil Market Report* (Paris: IEA/OECD, October, 2005); Energy Information Administration (EIA), *Short-Term Energy Outlook* (Washington, DC: USDOE, October, 2005); FACTS Inc. *Asia-Pacific Databook 1: Supply, Demand & Prices* (Honolulu, HI: FACTS, Fall 2005).

The global market has grown strongly since 1995, with incremental demand close to 1.5 million barrels per day (bpd) but dropped significantly in 1998 due to the Asian Financial Crisis. Just when the world seemed to have left the effects of the Asian Crisis behind, the relatively high oil prices of 2000-01 and a dramatic economic slowdown combined to dampen global oil consumption. The terrorist attacks of September 11, 2001 on the United States had lowered the oil prices in the final quarter of 2001 due to sluggish demand sentiment and its aftermath worsened an already weak demand outlook as its effects reverberated throughout the world.

Despite the outbreak of severe acute respiratory syndrome (SARS) in 2003, global oil demand recovered from the previous three years with well above the 1.5 million bpd level. An exceptional year followed in 2004, with 2.9 million bpd incremental demand, and the Asia-Pacific region accounting for 38.1 percent of this increase. Within Asia, China remains the key driver to demand growth and accounted for 71.3 percent of Asia's 1.1 million bpd incremental demand in 2004. This solid growth can be traced to China's robust economic performance with 9.5 percent real GDP growth and the corresponding growth in petroleum use in almost every economic sector.

Among the major oil consuming countries, India's demand growth is likely to remain strong due to expanding demand in the agricultural sector following robust growth over the past two years. Japan's demand has reverted to its declining trend, as the nuclear power issue that had driven its 2003 demand increase has been partly resolved. Although Japan's economic situation has improved recently with a mature economy and a soon-to-be shrinking population, its long-term oil demand growth prospects are not promising. South Korea has recovered somewhat from the brief recession of early 2003 caused by a consumer credit crisis, subsequent tightening of lending requirements and slowdown in household spending. It is expected to grow at a moderate rate as befits a maturing economy.

Other major consumers worthy of note include Thailand and Indonesia, where fixed or subsidized petroleum product prices are insulating consumers from relatively high oil prices and supporting demand growth. Although Thailand phased out its subsidies totally in July 2005, Indonesia has had a long history of substantial price subsidies and is struggling with the cost of maintaining these in the current high oil-price environment. In March 2005 the government raised the fuel prices by an average of 29 percent, and raised prices again by an average of 120 percent on October 1, 2005.

Over the last 15 years, the Big Four consumers in Asia – China, Japan, India, and Korea – have accounted for a little over 70 percent of total oil demand in the region. This is expected to remain roughly the same in the future, but the positions of the countries within the Big Four have changed recently (Table 8.2), where China took the top spot from Japan in 2003, and India took the third spot from Korea in 2004. However, the ranking of the current Big Four is not expected to change through 2015. Sustained high oil prices are likely to slow down the oil demand for some countries in the Asia-Pacific region, especially those with a deregulated market, which has left consumers exposed to the relatively high oil prices.

Table 8.2
Asia Pacific Petroleum Product Consumption by Country (2001-2004)

Country	Total Demand (kbpd)				Annual Growth (%)		
	2001	2002	2003	2004	2001–2002	2002–2003	2003–2004
Australia	780	791	793	809	10	3	15
Brunei	9	10	10	10	0.6	0.3	0.2
China	4680	4835	5316	6105	155	481	789
India	2082	2117	2171	2293	35	54	122
Indonesia	1096	1122	1139	1225	26	17	86
Japan	5218	5159	5243	5090	-59	84	-154
Malaysia	476	463	478	517	-13	15	39
New Zealand	127	139	149	146	13	9	-3
Pakistan	371	354	330	319	-17	-24	-11
Philippines	331	318	317	324	-13	0	7
Singapore	647	654	675	727	7	21	52
South Korea	2164	2217	2238	2218	53	21	-21
Taiwan	887	879	915	946	-8	36	31
Thailand	773	812	863	948	39	51	85
Vietnam	173	197	222	241	24	25	19
Others	577	584	597	645	7	13	48
Total	**20,391**	**20,651**	**21,456**	**22,562**	**260**	**805**	**1,106**

Source: FACTS Inc. *Asia-Pacific Databook 1: Supply, Demand & Prices* (Honolulu, HI: FACTS, Fall 2005).

As the Asia-Pacific region has been a primary driver of global incremental oil demand, any major slowdown in the region will have significant implications for the global oil market. In the longer term, as the Asia-Pacific economies become more developed, oil demand outlook is not likely to grow as spectacularly as it had in the early 1990s, but it should be growing at slightly better rates than it had been for the last rocky five years. However, even with slower growth, the region will still play a large role in global incremental demand, and is expected to account for about half of world demand growth in the coming decade.

Looking forward, the Asia-Pacific region as a whole is expected to see its consumption grow by a strong 700–800 kbpd annually through 2015. Although FACTS expects the growth of incremental demand to slow down as China's growth is tempered by relatively high oil prices and government intervention to cool the overheating economy, it projects that China's consumption will grow by a healthy 350–450 kbpd annually through 2015. China's logistical constraints, stemming from inadequate infrastructure, will contribute to a slowdown. However, this should not downplay the key role that China will continue to play in the global oil market. Under the FACTS base-case scenario, China's petroleum product demand is expected to reach 10.4 million bpd in 2015—nearly 4.0 million bpd higher than demand in 2004.

It is important to indicate that the growth scenario presented above is somewhat clouded by Japan, a major consumer stagnant with negative growth, which is unusual when compared to the rest of the region. South Korea's consumption growth is expected to be rather moderate with around 25–40 kbpd annually through 2015, while India is expected to grow more strongly with 80–110 kbpd annually for the same period.

The projections for Asia-Pacific petroleum product demand and annual oil demand growth are presented below (Figure 8.5). The projected average annual growth rates for 2005–2010 and 2010–2015 are 3.3 percent and 2.7 percent respectively. Among the Big Four consumers, China's share of Asian oil demand is expected to increase from 28.0 percent in 2005 to 33.0 percent in 2015; Japan's share is expected to decline from 22.0 percent to 16.0 percent for the corresponding years; India's share is expected to increase slightly from 10.0 percent to 11.0 percent; and South Korea's share is expected to decline slightly from 10.0 percent to 8.0 percent.

Figure 8.5
Asia-Pacific Petroleum Product Demand (1970–2015)

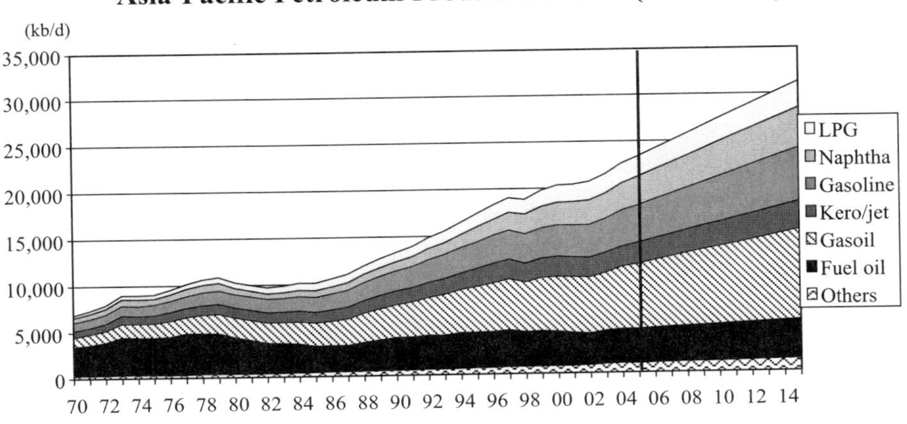

Source: FACTS Inc. *Asia-Pacific Databook 1: Supply, Demand & Prices* (Honolulu, HI: FACTS, Fall 2005).

Turning the focus to individual products, over the period 2005–2010, the consumption of naphtha (4.0 percent), gasoline (3.9 percent), and gasoil (3.6 percent) is projected to grow at or above the overall growth rate of regional petroleum product demand (3.3 percent). However, the consumption of LPG (3.0 percent), Kero/jet (3.2 percent) and fuel oil (1.2 percent) are projected to lag behind. Over the period 2010–2015, growth in the consumption of naphtha (3.0 percent), gasoline (3.4 percent), and gasoil (3.1 percent) are projected to exceed the overall growth rate (2.7 percent), which is brought down by lower growth in the demand for LPG (2.4 percent) and fuel oil (0.9 percent), with kero/jet (2.7 percent) growing at the same rate.

As the regional economies grow and people become more affluent, the consumption of transport fuels, such as gasoline and gasoil, is expected to grow faster than overall consumption due to an upsurge in car ownerships. Strong demand growth is expected for naphtha due to increased petrochemical activities, but LPG is projected to slow down somewhat from the extremely high growth seen in the past. It should be noted that although growth in fuel oil demand is relatively flat, the fact that it is projected to grow at all is somewhat unusual in comparison to other regions in the world, where it is generally expected to decline.

Outlook for Oil Production in the Asia-Pacific Region

In 2005 the Asia-Pacific region is expected to produce approximately 7.4 million bpd of crude, around the same as in 2004. The major producers in the region include China, Indonesia, Malaysia, India, Australia, Vietnam and Brunei. Under the FACTS base case scenario, regional crude oil production is projected to increase slightly to 7.5 million bpd in 2006, owing to moderate production increases in Australia, India and Malaysia. Regional production will increase to above 7.9 million bpd in 2008 before it declines slightly in 2010 due to dwindling output, particularly from Australia and India.

Over the last few years, the Big Six producers in the Asia-Pacific region – China, Indonesia, Malaysia, India, Australia and Vietnam – have accounted for a little over 92 percent of total oil production in the region and this is expected to remain roughly the same in the future. Unlike the major oil consumers in Asia, where the positions within the Big Four have changed rapidly over the past two years, there was no change in positions within the Big Six oil producers in the past few years, and the ranking is expected to remain the same through 2010.

The Asia-Pacific region's crude exports, mainly intra-regional exports, are estimated to continue to drop by 3.7 percent in 2005 after a decline of 9.4 percent in 2004. The drop of crude exports in 2005 is expected due to a likely decline of exports from Indonesia. With higher crude exports expected from Australia and a significant increase in Malaysia in the short term, FACTS projects that the regional crude export availability may increase to slightly over 1.9 million bpd in 2007, before declining to around 1.8 million bpd by 2010.

The Asia-Pacific region is known for its low sulfur sweet crudes despite a few streams containing sour crudes. These low-sulfur crudes may be divided into heavy (including medium) and light grades. Australia, Indonesia, Malaysia, Thailand, Pakistan, Papua New Guinea and some other smaller countries produce most of the light sweet crudes. China, Vietnam and others are heavy sweet crude producers while Indonesia, India and Brunei produce both. Overall, heavy sweet crudes (including medium grades) accounted for around 26 percent of the regional crude production in 2004, down from 28 percent in 2000. However, over the coming years, Malaysia and Australia may add some volumes of light sweet crudes, thus raising its share to as much as 30 percent in 2008 before settling at around 29 percent by 2010.

Rising Asian Oil Imports and the Role of Gulf

As a result of the rapidly growing oil demand and flat oil production, the Asia-Pacific region has a high ratio of oil import dependence. Of all the sources of oil supply, the Gulf remains the dominant and the most important one.

Outlook for Crude Oil Production and Exports of Major Gulf Countries

In 2004, the Gulf as a whole produced approximately 21.4 million bpd of crude and condensate. Saudi Arabia is by far the region's largest producer and exporter—accounting for 42 percent of the production from the Middle East—followed by Iran at 19 percent, and UAE, Kuwait and Iraq at approximately 11 percent, 10 percent and 10 percent, respectively (Figure 8.6). Qatar and Oman each account for 4 percent of the total production. Iraq is of particular interest, as the country has recovered to the pre-war level after a severe production decline in the aftermath of the March 2003 US-led invasion.

Figure 8.6
Key Middle Eastern Oil Producers

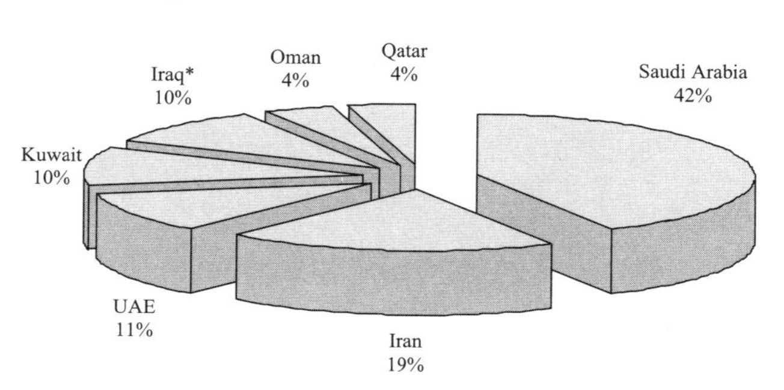

Notes: Based on estimated production in 2004 = 21.43 mmbpd

* crude and condensate; ** Saudi Arabia and Kuwait include their share of production in the Neutral Zone

Sources: FACTS Inc. *Middle East Petroleum Databook* (Honolulu, HI: FACTS, Fall 2005) and Organization of Petroleum Exporting Countries (OPEC). *Monthly Oil Market Report* (Vienna: October 2005).

Most Gulf producers have been producing close to capacity, with only small volumes of spare capacity—mostly in Saudi Arabia. Out of the total crude and condensate output in 2004, approximately 5.6 million bpd was consumed domestically in refineries and direct burning for power generation, and the remaining volumes of approximately 15.8 million bpd were exported. Of these exports, approximately 63 percent went to the Asia-Pacific region, followed by 16 percent to Europe, and 14 percent to North America. The remaining volumes were exported to Latin America and Africa. Although the Gulf supplies oil to virtually the whole world, the Asia-Pacific region has been the dominant buyer.

FACTS projects that production from the Gulf region will increase by about 4.4 million bpd between 2004 and 2010, from the 2004 output of 21.4 million bpd to an estimate of 25.8 million bpd by 2010. Much of these increases in output are expected to come from Saudi Arabia, Iraq, UAE and Kuwait. Exports from the Gulf region will increase from the availability of 15.8 million bpd in 2004 to an estimate of 18.4 million bpd by 2010.

Outlook for Asia-Pacific Imports of Crude and Refined Products

The Asia-Pacific region is heavily dependent on Middle Eastern crude for most of its import requirements. In 2004, almost three quarters of the 15.4 million bpd import requirements were fulfilled by the Gulf region (Figure 8.7). Intra-regional trade accounted for about 11.8 percent of the import requirements, and the remaining quantity was met by crudes from the Atlantic Basin, Africa, Americas and others. With domestic production projected to remain mostly flat, Asian imports will grow in the future on the back of rising demand—with imports of 18–19 million bpd becoming possible towards the end of the decade.

Figure 8.7
Crude Imports in the Asia-Pacific Region (2004)

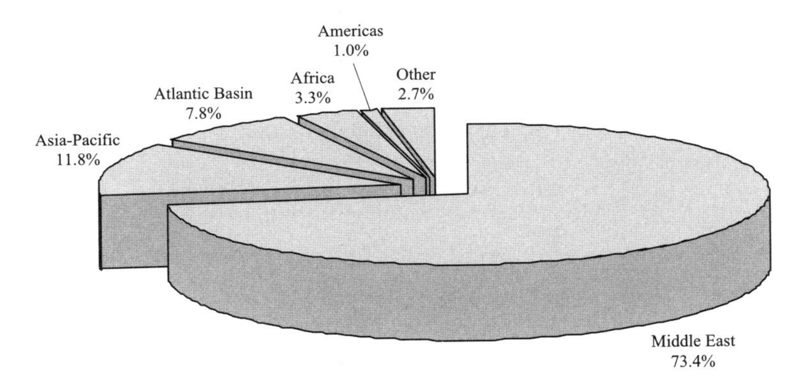

Notes: Estimate of import total = approximately 15.4mmbpd

Atlantic Basin = West Africa + Europe

Sources: FACTS Inc. *Asia-Pacific Databook 1: Supply, Demand & Prices* (Honolulu, HI: FACTS, Fall 2005); and FACTS Inc. *Middle East Petroleum Databook* (Honolulu, HI: FACTS, Fall 2005).

Much of the increase in imports will come from the Gulf, but Atlantic Basin crudes will also play an important role in some markets. The attractiveness of the Atlantic Basin crudes is their low sulfur content and relatively transparent prices compared to Middle Eastern crudes. For some refiners who are facing stagnant or declining domestic production of sweet crudes, the availability of sweet African/North Sea imports has permitted them to manage without substantial investment in refinery modifications.

Japan, China, South Korea, India and Taiwan constitute the top 5 crude importers in the Asia-Pacific region, and their combined imports accounted for more than 80 percent of the imports into the region in 2004. Japan had the heaviest dependence on Middle Eastern oil last year, where 88.9 percent of its 4.2 million bpd import requirements came from the Gulf. China is more diversified in its supply sources and only 45.5 percent of its 2.4 million bpd crude imports were of Middle Eastern origin. In contrast, South Korea, India and Taiwan depended fairly heavily on the Gulf, with between 70.3–78.1 percent of its import requirements coming from this region. The share of the Middle-Eastern crude imports to the Asia-Pacific region will increase from 72 percent in 2004 to about 75 percent by 2010 (Table 8.3).

Table 8.3
Asia-Pacific Crude Oil Imports (in kbpd) and Middle East Share

	2000	2001	2002	2003	2004	2010
Crude Runs	17,937	18,037	17,981	18,964	20,154	24,123
Crude Direct Use	360	310	293	342	338	328
Total Crude Demand	18,297	18,347	18,274	19,306	20,491	24,451
Asia-Pacific Crude Output	7501	7,419	7,518	7,342	7,392	7,852
Own Crude Use	5,254	5,269	5,350	5,269	5,441	6,085
Crude Imports	13,043	13,078	12,923	14,037	15,050	18,366
From within Region	1,847	1,778	1,942	1,809	1,742	1,711
From Africa/Americas/Europe	1,529	1,252	1,679	1,825	2,414	2,850
From Middle East	9,667	10,048	9,303	10,403	10,894	13,805
Middle East Share	**74%**	**77%**	**72%**	**74%**	**72%**	**75%**

Sources: FACTS Inc, *Asia-Pacific Databook 1: Supply, Demand & Prices*; *Asia-Pacific Databook 2: Refinery Configuration & Construction*; and *Asia-Pacific Databook 3: Oil Product Balances* (Honolulu, HI: FACTS, Fall 2005).

Focusing on refined products, it may be noted that the Asia-Pacific region's total refining output was way below the total product demand, thus requiring some 2.2 million bpd of imports in 1995 mostly from the Gulf. Total net imports into the Asia-Pacific region were relatively flat from 1995 to 2004, as refining capacities continued to be added to offset rising demand. FACTS expects Asia-Pacific regional refining capacity and production to continue to grow in the next few years. In the five years to 2010, Asia-Pacific regional capacity is expected to add another 3.67 million bpd—with 1.97 million bpd from China and 1.08 kbpd from India. However, the net imports, after these massive increases in refining capacities, are expected to increase steadily from 2.1 million bpd in 2005 to reach about 2.4 million bpd in 2010.

The Role of the Gulf in the Asia-Pacific Region

Crude oil markets in the Asia-Pacific region are characterized by flat regional crude production, declining crude export availability and rising crude imports. Overall, Middle Eastern crude oil accounts for over 70 percent of the region's imports. At the same time, the Asia-Pacific region is the largest customer of the Gulf, accounting for over 60 percent of the crude exported from the Middle East. Furthermore, the geographical proximity of the two regions, a growing Asian deficit and the absence of alternative sources have increased the dependency of Asian countries such as Japan, South Korea, India, China and Taiwan on the supply of Middle Eastern crudes.

The Asia-Pacific region and the Middle East are not only closely intertwined in terms of their crude trade, but also in their petroleum product markets. Besides their proximity, the Asia-Pacific and Middle Eastern petroleum product markets are closely linked due to a lack of viable alternatives, as much of the surplus petroleum products from the Gulf cannot meet the more stringent product specifications in the

European or North American markets, and therefore have to be shipped to Asia. Yet, it should be mentioned that many of the new refineries planned in the Middle East will be able to meet the product specifications in the West as well. For the time being, Latin America provides one of the few remaining markets where specifications are relatively lax, although its capacity to absorb imports is limited.

FACTS expects that demand in the Asia-Pacific region will continue to grow and based on the planned additional refining capacity, the product market will continue to be tight. However, the expected refining capacity additions from the Middle East also imply that there will be an increase in product export availability from 2.9 millon bpd in 2005 to 4.6 million bpd by 2010. FACTS recognizes that LPG, naphtha and fuel oil have dominated Asia-Pacific net product imports and will continue to do so through 2010. Although the region depends heavily on LPG imports, about 80 percent of these imports are sourced from gas separation plants in the Gulf. As such, there is no tightness in the refining sector caused by the massive imports of LPG. Furthermore, the Gulf is expected to have more than enough to meet the requirements of the Asia-Pacific.

However, the product balances for naphtha are quite different, with only 37 percent of Gulf output being supplied from non-refinery sources. Furthermore, the naphtha surplus from the Gulf is insufficient to meet the Asia-Pacific deficit, which has to be met by other regions. Gulf fuel oil exports are also inadequate to meet the region's requirements. These trends will bolster regional product prices and help create market tightness. However, in the future, Middle Eastern exporters will find it more difficult to sell their other products, such as gasoline, gasoil and kero/jet in the Asia-Pacific region, as Asia is expected to continue to produce more than it requires. However, it is expected that a major portion of surplus products from the newer refineries will go westward, such as gasoil to the European market. Major upgrading projects will be required for many existing refineries in the Gulf if they are to meet the more stringent specifications prevailing in the West.

In summary, given current trends in production and consumption in both crude oil and refined products, it is clear that these two regions will become increasingly linked in the future. The Asia-Pacific with its rising demand will constitute an even larger share of Middle Eastern crude due to increasing export availability in the Gulf. However, based on FACTS' projected petroleum product balances for the Middle East and Asia-Pacific, only imports of LPG, naphtha and fuel oil are likely from the Gulf to the Asia-Pacific region, with surpluses in other products. The increasing level of oil dependence on the Gulf is likely to result in close economic and political ties between the Asia-Pacific and the Gulf regions.

China's Growing Demand and Imports: Role of the Gulf

China is currently the world's second largest primary energy consumer after the United States. In 2004, China consumed over 27.4 million boepd of PCEC while the United States consumed nearly 47 million boepd. In Asia, China is the largest consumer of oil and coal as well as primary commercial energy as a whole. The impact of China's growing energy demand and imports go beyond just the Asia-Pacific region and can be felt by the Gulf too. In this section, various issues related to China's rising energy demand and import dependence are addressed.

China's Growing Energy Demand and Changing Structure

During the past quarter century, China's PCEC growth was slower than its official GDP, but still impressive. During the 1980–2004 period, the PCEC grew at an average annual rate of 5.3 percent, while the average annual growth rate for real GDP was 9.5 percent. For 2003 and 2004 in particular, China had double digit PCEC growth. The total PCEC reached 27.4 million boepd in 2004, up from 8.0 million boepd in 1980, 12.7 million boepd in 1990, and 18.1 million boepd in 2000 (Figure 8.8).

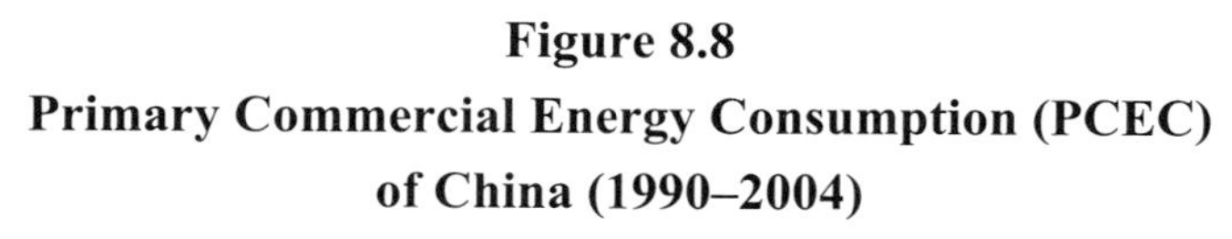

Figure 8.8
Primary Commercial Energy Consumption (PCEC) of China (1990–2004)

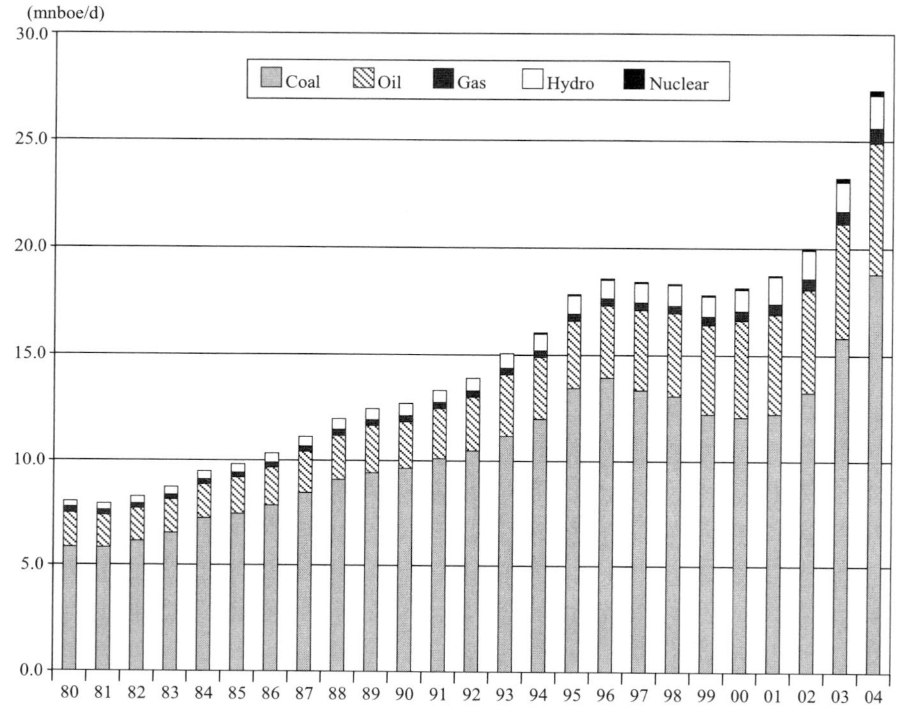

Sources: China National Bureau of Statistics (CNBS), *Statistical Yearbook of China* (Beijing: CNBS, 2005); and FACTS Energy Database.

China's PCEC is dominated by coal. In 2004, coal accounted for 68.7 percent of the PCEC, down from 73 percent in 1980 and 76 percent in 1990, but up from 66 percent in 2000. Currently coal is responsible for 75–80 percent of China's power generation. Over the next ten years, coal use in China is forecast to continue, though its share in total PCEC is likely to decline notably.

China has surpassed Japan and become the largest oil consumer in Asia. In 2004, China consumed 6.1 million bpd of petroleum products (including direct burning of crude oil). The oil share in China's PCEC was

22.3 percent in 2004. Over the next ten to fifteen years, China's oil demand is poised to grow strongly, and its share in the country's total PCEC is expected to increase slightly. More discussions relating to the oil demand, supply and trade will be provided in the following sections.

Natural gas has been underutilized in China. Its share in the country's total PCEC is low—under 3 percent in 2004. However, the future use of natural gas in China is expected to grow rapidly. There are three sources of growth on the supply side: rising domestic gas production, incoming LNG, and future imports of pipeline natural gas. Unlike oil, China's domestic gas output has been growing fast though it started from a small base. For LNG imports, two regasification terminals are being built: one is in Guangdong Province and one in Fujian Province, which have a combined capacity of over 6.4 million metric tons per year. Beyond these two, more terminals are being planned along the coastal area. For future LNG imports, China is not only targeting Asian suppliers, but also the Gulf suppliers—Qatar, Iran and Yemen. For international pipeline gas, the likely sources are the Russian Far East, East Siberia and West Siberia. Over the long term, China may also seek to import natural gas from Central Asia. Taking all the above aspects into consideration, China's natural gas consumption is expected to grow rapidly and its share in China's total PCEC is likely to rise significantly, though from a low starting point.

Hydroelectricity plays an important role in China's power generation and energy supply. At the start of 2005, China had over 100 gigawatts (GW) of installed hydroelectric power generating capacity. Currently, the huge Three Gorges Hydroelectric Power Plant is partially operational. By the time all generators are installed by 2008-09, the total capacity of the Three Gorges Plant will reach 18.2 GW. In the meantime, China plans to build more dam projects in the southwest and other parts of the country. FACTS expects that by 2020, the share of hydroelectricity in China's total PCEC will be higher.

China is a late comer in nuclear power development. Although the first reactor generation unit was installed in 1991, development remained slow during most of the 1990s. Even in the early 2000s, the buildup of nuclear power was cautious. However, since 2003-04, as China's energy and power demand grew rapidly, the Chinese government has accelerated the nuclear power program and vowed to make it a credible contributor to the country's overall power supply. As of mid-2005, China had roughly 8 GW of installed nuclear power generating capacity from nine units. More reactor generators are being built. The government's target is to raise the nuclear power generating capacity to 40 GW by 2020. If that happens, the share of nuclear power in China's total PCEC will be significantly higher than the 2004 level of below 1 percent.

Meeting China's Rising Oil Demand: Role of the Gulf

China's oil demand has been growing rapidly, while the country's domestic oil production has been growing slowly. This has resulted in rising oil imports since the early 1990s. In the meantime, China's refining sector has been expanded vigorously. Consequently, the majority of China's oil imports is crude oil. This pattern is expected to continue for the coming years.

China's oil production growth slowed down considerably in the 1980s, followed by further stagnation in the 1990s. For instance, the average annual growth rate of China's domestic crude production was 19.4 percent in the 1960s and 13.2 percent in the 1970s, but it declined to 2.7 percent in the 1980s and 1.6 percent in the 1990s. In 2004, China produced 3.5 million bpd of crude oil, up from 2.1 million bpd in 1980, 2.8 million bpd in 1990, and 3.2 million bpd in 2000 (Figure 8.9). The high oil prices did provide an impetus for China's oil production since 2004. For 2005, China is likely to see 4 percent growth in crude oil production, reaching 3.6 million bpd.

Figure 8.9
China's Crude Oil Production (1978-2004)

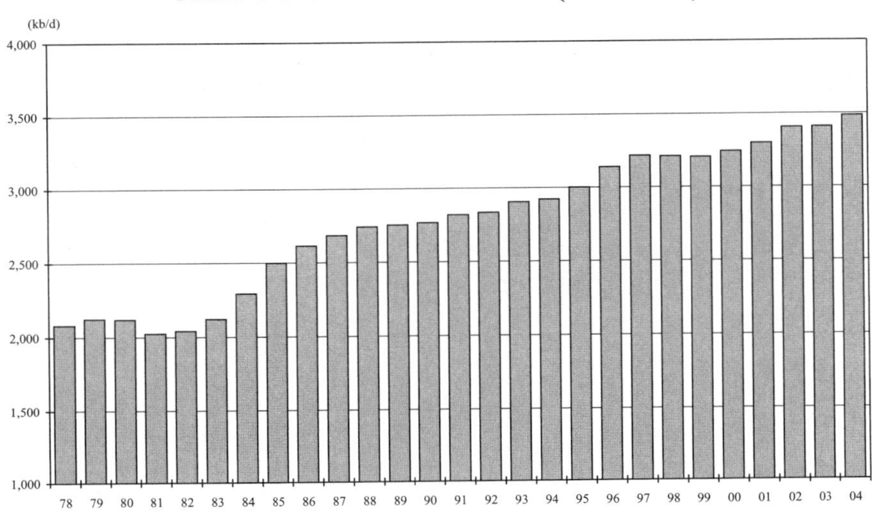

Sources: China National Bureau of Statistics (CNBS), *Statistical Yearbook of China* (Beijing: CNBS, 2005); and FACTS Energy Database.

At the start of 2005, China's total crude distillation capacity was estimated at 6.5 million bpd (although around 350-400 kbpd of small locally-owned refineries cannot be fully identified), up from 3.1 million bpd at beginning of the 1990s. China Petrochemical Corporation (Sinopec) accounts for 53 percent of the existing refining capacity in China, followed by China National Petroleum Corporation (CNPC) and its public listed subsidiary PetroChina at 41 percent, and others at 6 percent. In 2004, crude runs in China reached 5.4 million bpd, up substantially from 1.6 million bpd in 1980, 2.2 million bpd in 1990, and 4.2 million bpd in 2000 (Figure 8.10).

As far as configuration is concerned, the Chinese refining sector exhibits the following characteristics: [4]

- Large overall crude distillation capacity but most refineries are of small size
- High fluid catalytic cracking (FCC) and residual catalytic cracking (RCC) capacity

- Uneven distribution of refining facilities among different regions
- Low catalytic reforming capacity
- Low utilization rate though it has been improved since the early 2000s. The situation is more severe for CNPC/PetroChina since markets within its geographical area are much smaller than that of Sinopec.

Figure 8.10
China's Refining Throughput (1990–2004)

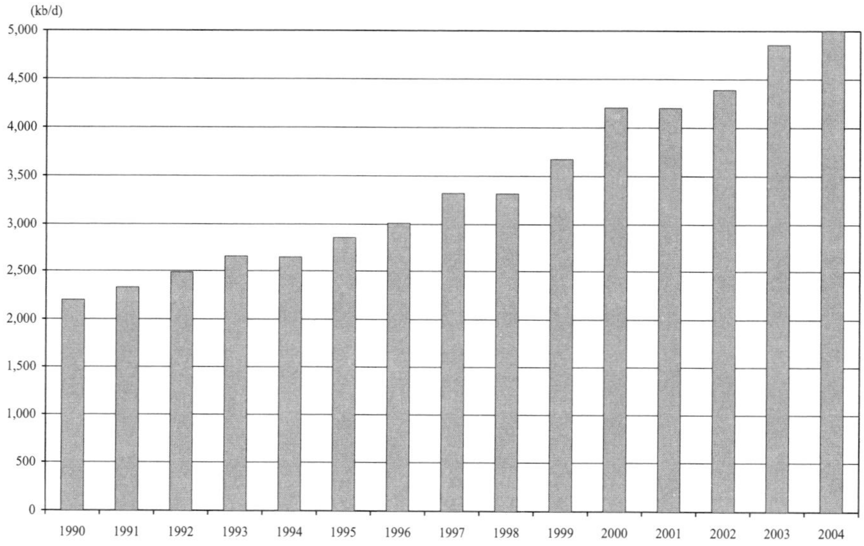

Source: FACTS Energy Database.

In the meantime, however, China is under pressure to increase its sour-crude processing capacity to deal with increasing Gulf crude imports and declining availability of sweet crudes within the Asia-Pacific region. Under this circumstance, China's total sour crude (1 percent sulfur or higher) handling capacity reached 1.3 million bpd at the start of 2005, and it is expected to increase further during the coming years. Sinopec accounts for around 80 percent of China's sour crude handling capacity and the rest are in the hands of CNPC/PetroChina.

Petroleum product demand in China is characterized by spectacular growth—especially since the early 1990s—and a radical transformation of the demand pattern. Largest in the Asia-Pacific region right after Japan, China's petroleum product demand reached 6.1 million bpd in 2004—including direct use of crude oil in industrial sectors and for power generation (Figure 8.11). During the past two plus decades (1980–2004), petroleum product demand growth averaged 5.7 percent per annum, in which growth accelerated to 7.5 percent per year on average since the 1990s. In 2004, China's petroleum product demand growth reached nearly 15 percent.

Figure 8.11
China's Petroleum Product Consumption (1980–2004)

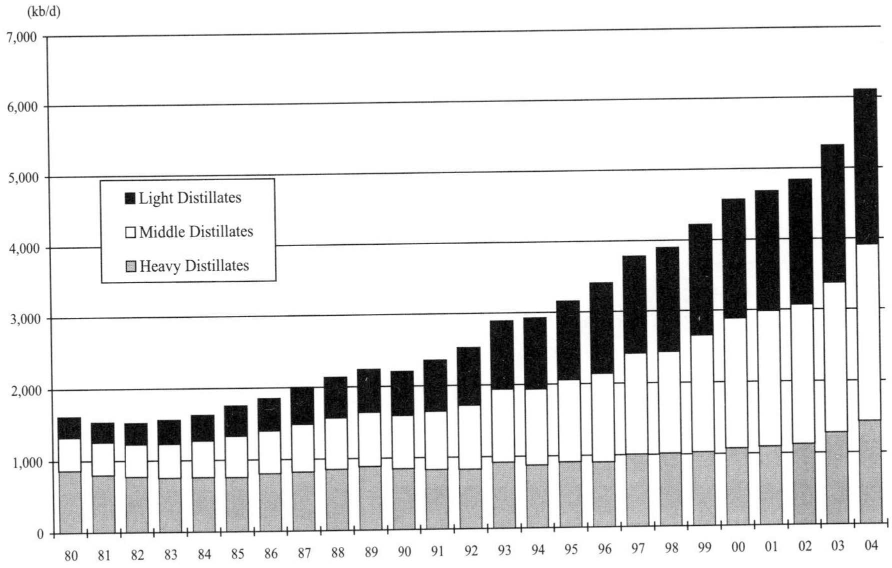

Source: FACTS Inc., *Asia-Pacific Databook 1: Supply, Demand & Prices* (Honolulu, HI, Fall 2005).

While total petroleum product growth remained strong in China during the 1990s, the same cannot be said about individual products. China's product demand pattern has undergone a radical transformation in response to its past economic and energy policies. China's product demand slate in the early

1980s was heavily oriented toward bottom-of-the-barrel products, with heavy distillates (fuel oil, crude oil for direct burning, and other products) accounting for 43 percent of the total demand in 1985. This share has steadily dropped, owing to the government's policy of minimization and eventual substitution of fuel oil use in power plants by coal. In 2004, heavy distillates constituted 23 percent of China's total product demand. As the share of heavy distillates declines, the share of light-to-medium distillates rises. The strong demand growth for transportation fuels and for feedstock to serve China's growing petrochemical production capacity, has also contributed to lightening the country's product demand barrel.

China's crude and product exports peaked in the mid-1980s at 600 kbpd (thousand barrels per day) and 125 kbpd, respectively, but have since declined. Meanwhile, imports of crude and to a lesser extent, products have increased rapidly. China has become a net overall oil importer since 1993. In 2000, net imports reached just under 1.5 million bpd and in 2004, touched an all-time high of over 3 million bpd (Figure 8.12).

Figure 8.12
China's Oil Exports and Imports (1980–2004)

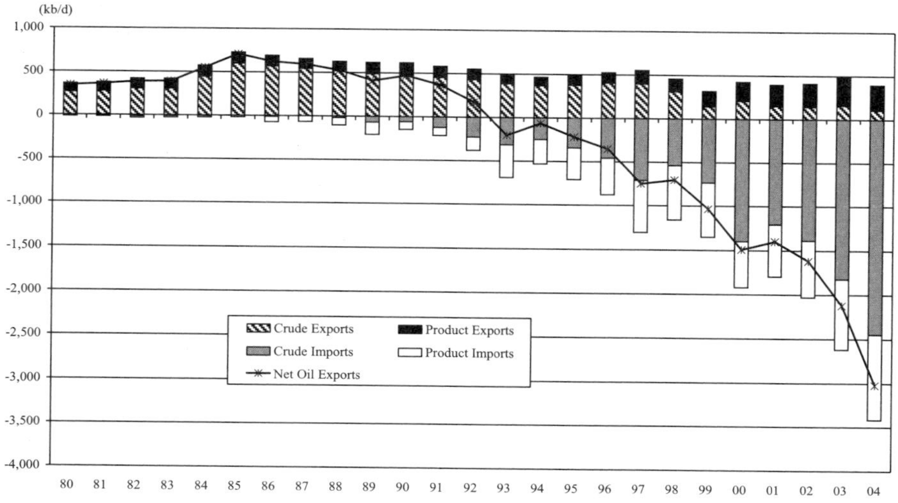

(kb/d)

Crude Exports
Product Exports
Crude Imports
Product Imports
Net Oil Exports

Source: Kang Wu and Fereidun Fesharaki. "As Oil Demand Surges, China Adds and Expands Refineries," *Oil and Gas Journal* vol. 103, no. 28 (July 25, 2005).

China is still a crude exporter, but the volume is much smaller today than a decade ago. In 2004, China's crude exports were 100 kbpd, down from 601 kbpd in 1985 and 377 kbpd in 1995. Prior to 2003, Japan had long been the leading importer of Chinese crude. With the ending of the government-to-government export agreement for Daqing crude between Japan and China in early 2004, China's total crude exports have declined and so has the share of Japan. In 2004, Japan was overtaken by South Korea, Indonesia, the United States, North Korea and Malaysia and became only the fifth largest crude oil importer from China, averaging about 10 kbpd (Table 8.4). During the first ten months of 2005, Japan's imports of Chinese crude rose to 16 kbpd, but the country remained behind South Korea, the United States and Indonesia.

Table 8.4

China's Crude Exports by Destination (2004)

Destination	Volume (kbpd)	Share (%)
South Korea	28.8	26.3
Indonesia	27.6	25.2
USA	11.6	10.6
North Korea	10.6	9.7
Malaysia	9.9	9.1
Japan	9.8	9.0
Australia	8.9	8.1
Others	2.3	2.1
Total	**109.5**	**100.0**

Source: China General Customs Administration.

Saudi Arabia topped the list as the leading crude exporter to China in 2004, followed by Oman, Angola, Iran, Russia, Sudan and others (Table 8.5). During the first ten months of 2005, Saudi Arabia continued to top the list of crude exporters to China, followed by Angola, Iran, Russia, Oman, Sudan, Yemen and others. The rise of Saudi Arabia, Iran and other

Gulf countries in China's crude imports indicates the rising sour crude handling capacity in the country as well as the increasing importance of the Gulf oil supply to China.

Table 8.5

China's Crude Imports by Source (2004)

Source	Volume (kbpd)	Share (%)	Source	Volume (kbpd)	Share (%)
Saudi Arabia	343.9	14.0	Nigeria	29.7	1.2
Oman	326.1	13.3	Libya	26.7	1.1
Iran	264.0	10.8	Chad	16.6	0.7
Yemen	98.0	4.0	Algeria	13.5	0.6
UAE	26.8	1.1	Gabon	10.9	0.4
Iraq	26.1	1.1	Cameroon	2.6	0.1
Kuwait	25.0	1.0	Others	1.0	0.0
Qatar	2.8	0.1	*Africa Total*	*704.1*	*28.7*
Middle East Total	*1,112.7*	*45.4*	Russia	214.9	8.8
Vietnam	106.7	4.4	Norway	40.1	1.6
Indonesia	68.4	2.8	Brazil	31.5	1.3
Malaysia	33.7	1.4	Kazakhstan	25.6	1.0
Australia	30.1	1.2	Argentina	14.2	0.6
Thailand	18.2	0.7	Venezuela	6.7	0.3
Brunei	17.6	0.7	Ecuador	5.6	0.2
Others	7.7	0.3	UK	3.1	0.1
Asia-Pacific Total	*282.5*	*11.5*	Azerbaijan	2.6	0.1
Angola	323.3	13.2	Georgia	1.6	0.1
Sudan	115.1	4.7	Others	4.4	0.2
Congo	95.2	3.9	*Europe/Other Total*	*350.3*	*14.3*
Eq. Guinea	69.5	2.8	**Total**	**2,449.6**	**100.0**

Source: China General Customs Administration.

The product trade in China has been undergoing some dramatic changes since the early 1980s. Prior to the policy shift that relaxed petroleum product import restrictions in 1986, China's product consumption was supplied almost entirely through domestic production. During the period 1980–1985, China exported an average of about 110 kbpd of refined products and imported just under 30 kbpd each year. Since then, China's product imports have risen from a low of 18 kbpd in 1986 to 967 kbpd in 2004, though in 2005 the imports are expected to lower.

China's petroleum sector and oil markets are expected to undergo continuous change over the next ten to fifteen years. On the supply side, crude production growth from within China is expected to be flat. For petroleum product demand, the growth is likely to be strong. The results are a continuously rising import requirement for oil over the long term.

According to FACTS,[5] the total oil consumption (petroleum product demand plus direct use of crude oil) in China is forecast to grow at an average annual rate of 5.0 percent during the period 2004–2015 and is projected to reach 8.5 million bpd in 2010 and 10.4 million bpd by 2015 under the FACTS base-case scenario (Figure 8.13).These projections are very sensitive to alternative assumptions under different scenarios.

On the supply side, China's upstream oil industry faces a precarious situation, as production from Daqing and Shengli oil fields is stagnating, Huabei oil production is declining, and Liaohe field production is increasing only slowly. The hope for incremental production is likely to come from the West, the Northwest and offshore fields. On an overall basis, China's crude production is projected to grow steadily but slowly, reaching 3.7 million bpd in 2010 and 3.8 million bpd by 2015. As domestic production continues to lag behind demand, China's net oil import requirements (including both oil and products) are expected to reach 5.3 million bpd in 2010, and 6.8 million bpd by 2015 (Figure 8.14).

Figure 8.13
China's Petroleum Product Demand (2000–2015)

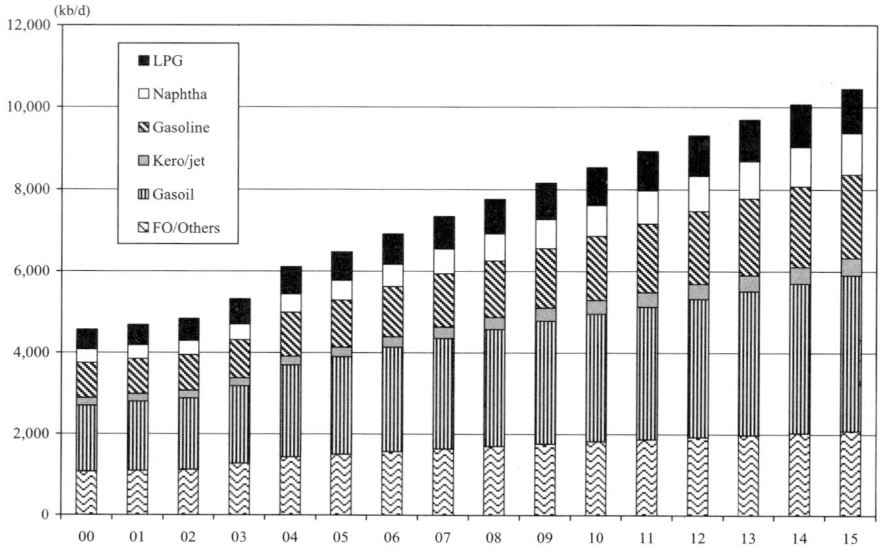

Source: FACTS Inc. *Asia-Pacific Databook 1: Supply, Demand & Prices* (Honolulu, HI: FACTS, Fall 2005).

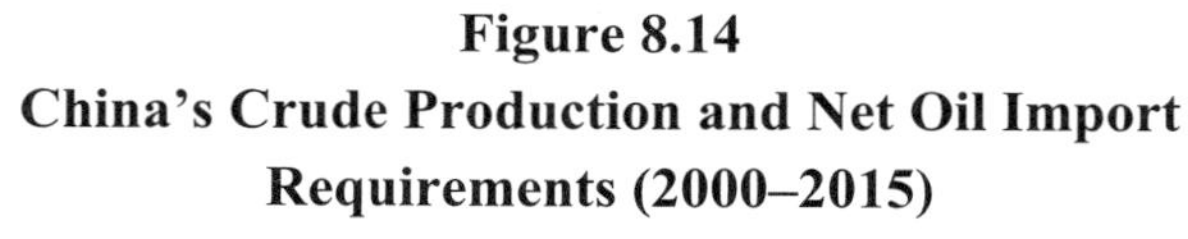

Figure 8.14
China's Crude Production and Net Oil Import Requirements (2000–2015)

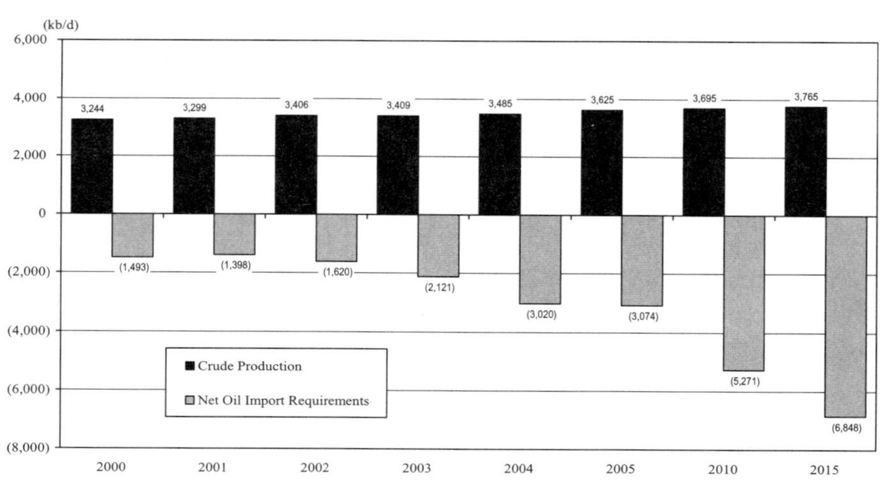

Source: Authors' forecasts

In view of these imports, the role of the Gulf, which is already important, will rise steadily. In 2004, the Gulf accounted for 45 percent of China's crude oil imports. The share is projected to reach 61 percent in 2010 with the increase of overall crude imports by China. If refined products are added, the share of the Gulf in China's overall oil import is even higher. With its huge additional oil import requirements, China's rising dependence on the Gulf constitutes one of the most important developments in Asia for the next decade and beyond.

Concluding Remarks

This chapter has highlighted the rising impact of the Asia-Pacific region on the global oil markets, and the significant role of the Gulf in meeting the growing Asian oil demand in general and Chinese demand in particular. While the region's oil demand is growing fast its crude supply has lagged behind. As a result, crude imports into the region are likely to increase, with much of the future increase expected to come from the Gulf. Due to massive refining capacity additions in the coming years, the Asia-Pacific region will have surpluses in most products except for LPG, naphtha and fuel oil, much of which are also expected to come from the Gulf. The Asia-Pacific region's petroleum product demand will not expand at rates that will absorb much of incremental output from the Gulf and more upgrading projects are bound to be launched to meet the more stringent specifications in the West. China is important for shaping the future growth of oil demand in the Asia-Pacific region. The country is a major force behind the region's move to higher levels of oil requirements and deepening import dependence on the Gulf. Overall, the increasing level of Asian oil dependence on the Gulf is likely to foster a closer economic and political relationship between the two regions.

9

Gulf Oil in a Global Context: Strategies for and Challenges to Demand Security

*Peter R. Odell**

In order to interpret strategies and identify challenges relating to the future demand for Gulf oil, it is appropriate to begin with developments in the mid-20th century. By then, the most important elements determining the prospects for Gulf oil had already been established, following the exhilarating and complex set of events and developments that had occurred in that newly emerging oil region during the years immediately following World War II.

Between the years 1945 and 1960, the Gulf region, including contiguous areas of Iran and Iraq, was revealed as an area of extraordinary and possibly unique oil resources. Hints of such a phenomenon had emerged even by the 1930s, but confirmation of the situation was delayed by wartime conditions.[1] Thus, in 1945, Gulf oil reserves were declared at a mere 3000 million tons, while the oil production that year was only 50 million tons, coming largely from Iran and Iraq and constituting less than 10 percent of global production.

By 1960, production had increased ten-fold and reached 25 percent of world output. Likewise, declared oil reserves had grown enormously to almost 25,000 million tons, with the western Gulf states, rather than Iran and Iraq, accounting for most of the massive increase. The region's reserves were now six times those of both the United States and the Soviet Union and they also gave the Middle East a reserves-to-production (R/P) ratio of nearly 100 years.[2] Thus, the international community at that time had no difficulty in accepting the idea of "oil in the Gulf" as a promising situation. Subsequently, however, this optimism has been undermined by a series of restraints on oil production and exports as detailed below.

Gulf Oil Industry: Structural Antecedents

Over the same period – and in the context of their much greater knowledge of the actual and potential wealth of the Middle East – all of the world's eight major oil companies, viz. the "seven sisters" plus the CFP, participated in the frenetic action to secure concessions: nominally in competition with each other, but in practice, usually carving up the countries and their oil-rich areas into negotiated "interests." This process not only reflected their relative strengths as companies, but also the contrasting political and diplomatic weights of the United States, the United Kingdom and France.[3]

As the late Professor Edith Penrose demonstrated in her seminal late-1960s work, *The Large International Firm in Developing Countries*,[4] these companies exercised joint control over most of the reserves and production of the region. Their complex, inter-weaving linkages in the Middle East's upstream oil industry at that time are shown in Fig.9.1. The international community also readily perceived the critical role and influence of these companies in determining the future prospects of the region's oil.

The combination of the above-mentioned phenomena – the huge volumes of available oil; the range of countries under which it lay; the eight international petroleum corporations with regional oil concessions and finally, the political and economic interests of their mother countries – came to be widely perceived as having generated a potential "unholy" alliance, wielding a concentrated and critical mass of politico-economic power.[5] The dangers of this alliance were already clear and, indeed, perceived as getting worse in the global context of rapidly increasing oil usage, largely at the expense of indigenous energy, notably coal. Concern for undue dependence on Gulf oil was thus already established a long time ago.[6]

Figure 9.1
Ownership Links between Major International Oil Companies and Major Crude Oil Producing Companies in the Middle East (1966)

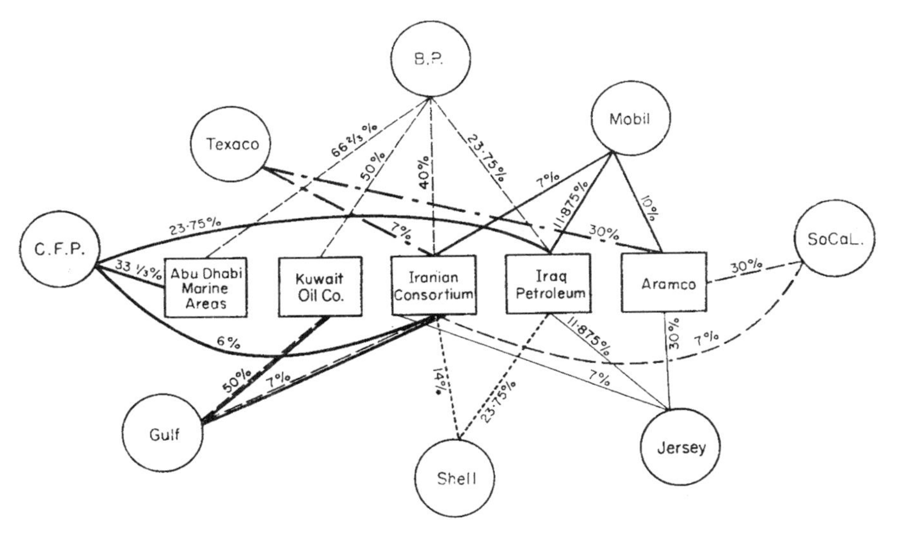

Source: E.T. Penrose, *The Large International Firm in Developing Countries* (London: Allen and Unwin, 1968).

Insecurities of the Gulf's Oil Reserves and Production

Since 1960, concern over undue dependence on Gulf oil has been more or less continuously exacerbated by security issues.

Cold War fears of regional intervention by the Soviet Union: Soviet intervention in the region could be prompted by either of two factors: first, as argued over many years, from a perceived Soviet need for access to Arabian Gulf oil, to sustain its expanding and increasingly energy-intensive economy; or second, as argued somewhat later, from fears of exposing the economy of the Western world to the Soviet Union's potential ability to deny it access to Arabian Gulf oil supplies.[7] Even the Soviet invasion of Afghanistan as late as 1980 was interpreted as a move to achieve that objective.

- *Concern about the stability of individual Arabian Gulf states:* Such concerns about the internal stability and viability of oil-producing states in the region have been raised in several cases: Iran in the 1950s, Saudi Arabia in the 1970s and Iraq since 1990. These concerns have materialized from time to time, leading to interruptions in oil flows and the renunciation of oil concessions with both kinds of events raising the risk profiles of the countries involved.
- *Recognition of the potential for the use of oil as a weapon*: Since the Suez Crisis of 1956–57, the potential to employ oil as a weapon has been a matter of concern, particularly in the context of the conflict between Arab countries and Israel, given that the former group views the existence of the latter as a consequence of Western actions or inaction.[8]

These negative perceptions regarding the security of Middle East oil reserves and the reliability of supplies from the region have proved to be more than conceptual and theoretical. On the contrary, they have been justified by interruptions to exploration and exploitation in general, and to supplies in particular, on many occasions since 1950—for example, from 1951 to 1954 due to a revolution in Iran; in 1956–57, 1963 and 1967 over the issue of Israel; in the late 1960s over western domination in the exploitation of Iraqi oil; in 1973–74 with the Arab countries' embargo on supplies to the industrialized importing countries; in 1979-80 from Iran's second revolution; in the early to late 1980s from the Iraqi–Iran War; in 1990–91 as a result of the Gulf War, in the 1990s by the United Nation's embargo on Iraqi oil and most recently, from the War in Iraq.[9]

Quite apart from these political and military reasons, there have also been supply constraints in the interests of upholding the price. In the 1950s and 1960s such a situation arose because of agreement by the operating companies on production levels[10] and in more recent decades, due to the influence of Gulf oil producers on OPEC policy-making, notably with regard to quota limitations.[11]

The demise, first, of the companies' joint control over oil supply and pricing mechanisms and second, of their "ownership" of most of the region's oil reserves and resources in the 1970s, has not led – except occasionally and for short periods – to an uncontrolled flow of oil from the region's prolific reserves and its well-developed production and transportation infrastructure. Individual countries inhibited production at different times as a matter of policy, but in addition, various combinations of countries have done the same, sometimes acting as allies, but more often as rivals. Beyond this, collective decisions by all or most of the oil producing countries of the Gulf have often kept supplies off the market. Indeed, such decisions have been exercised over a more extended period than that in which they have permitted their oil to flow freely in a competitive market in which prices would have declined towards the long run low supply price of oil.[12] Overall, the world's experience of volume restraint on the Gulf's oil production and exports in over 40 years has served, more or less continuously, to reinforce the international community's perception of the dangers stemming from dependence on oil supply from the region.[13]

The Gulf's Changing Role in Global Oil Supply and Trade

Role of the Gulf from 1955–1985

The evolution of the Arabian Gulf's contribution to global oil supply and trade over the period since 1955 is indicated in Table 9.1.

Over the first 20 years to 1975, the oil companies had a high degree of control over Gulf oil, but this was never absolute. This situation ensured a rising exploitation of regional reserves, so that annual production during that period rose six-fold, from 165 to 975 million tons. Over this period, the region's contribution to international oil trade rose from just under 50 percent of the 291 million tons traded in 1955, to over 60 percent of the 1508 million tons traded in 1975, constituting a more than 600 percent increase in the volume of Arabian Gulf oil moving into world markets. In essence, the international oil corporations exploited Gulf oil to the degree that it suited their strategies and enhanced their global after-tax profits, particularly after taxes paid to producer governments were made allowable against taxes at home.[14] Even so, by 1975 the region's reserves-to-production ratio (R/P ratio) had only been reduced from its level of nearly 100 years to a still more than adequate 50 years.

After 1975, when state action had eliminated much of the oil companies' equity interests in the Gulf's upstream oil, the region's production and its contribution to world supplies began to decline. Eventually, by 1985, it fell to little more than half of their 1975 peak values.[15] In the course of this decade, a near-stagnant global oil industry effectively turned its back on Gulf oil supplies in the context of regional uncertainties. These uncertainties were created by a sequence of events arising from the new national ownership and control of oil supply and prices; the Iranian Revolution in 1979; and the political and military disruptions caused by the Iran–Iraq War in the 1980s. The whole region, in general, was also full of fears for the future and was adversely affected by the impact of the Soviet Union's intentions in Afghanistan. The international community's adverse perceptions of the future of Gulf oil had materialized with a vengeance.[16]

Role of the Gulf from 1985–2005

Initially, following the 1986 oil price collapse and the re-emergence of growth in global oil demand, Gulf oil supply boomed. By 1990, output was up by two-thirds from its 1985 low, while its exports grew by almost 60 percent. It seemed as though the attractions of the Gulf's available production and transportation capacity and of competitive prices in a low-cost environment were successfully overcoming the negative international perceptions of renewed dependence on the region's oil industry.

Table 9.1

Global, Gulf and Regional Oil Production and Trade from 1955 to 2005 (in millions of tons)

	1955	1960	1965	1970	1975	1980	1985	1990	1995	2000	2005
1. Global Oil Production	786	1079	1450	2352	2734	3082	2790	3179	3282	3614	3895
of which from the Gulf	164	264	385	689	975	927	514	862	951	1102	1163
- as a % of Global Production	20.8%	24.2%	26.5%	29.3%	35.7%	30.1%	18.4%	27.1%	28.9%	30.5%	29.9%
2. Oil Traded Internationally	291	456	677	1263	1508	1588	1264	1612	1815	2112	2462
- as a % of total production	37.0%	42.3%	46.7%	53.7%	55.2%	51.4%	45.3%	51.2%	55.3%	55.8%	63.2%
3. Gulf Exports	145	229	340	631	918	864	447	704	825	942	982
- as a % of internationally traded oil	49.8%	50.2%	50.2%	50.0%	60.9%	54.7%	35.4%	43.7%	45.5%	44.6%	39.9%
4. Intra-Regional Trade:											
- Total	106	190	238	527	402	433	539	570	643	719	936
- as a % of international trade	36.4%	41.7%	35.2%	41.7%	26.0%	27.3%	42.6%	35.4%	35.4%	33.6%	38.0%
- as a % of Gulf exports	73.1%	83.0%	70.0%	83.5%	43.7%	50.1%	120.6%	81.0%	77.6%	76.3%	95.3%
- of which:											
a. Western Hemisphere	87	129	124	180	177	168	182	199	267	304	378
b. Europe/W Africa	11	42	158	310	175	193	308	313	296	294	429
c. East Asia/Australia	8	9	7	37	50	72	49	58	80	121	129

Sources: Data from relevant issues of BP's Annual *Statistical Review of the World Oil Industry from 1955 to 1980* and its *Statistical Review of World Energy from 1985 to 2006.*

However, this was abruptly cut short by the 1990–91 traumas of Iraq's takeover of Kuwait, its threat to the Saudi Arabian oilfields and the consequential Gulf War. Nevertheless, in the decade from 1990 to 2000, the Gulf did improve its position in the global oil market: with an increase in production of 24.6 percent compared with a 13.7 percent overall global increase; and a 33.8 percent growth in exports compared with a 27.5 percent overall expansion in global oil trade. However, since 2000, the comparable figures for growth in oil production are 5.5 percent for Gulf oil and 7.8 percent for global oil; and for growth in international trade, 4.2 percent and 16.5 percent respectively.

Such data confirms the hypothesis that, after the oil crises and revolutions of the 1970s and early 1980s and the subsequent wars in the region, the Gulf has been constantly struggling to reassert its international status. Its share of global production – standing at 29.9 percent in 2005 – is still well behind its 35.7 percent in 1975. Even more emphatically, its share of world oil trade remains stuck in the range of 40 to 45 percent (43.7 percent in 1990; 44.5 percent in 1995; 44.6 percent in 2000 and 39.9 percent in 2005). These statistics compare unfavorably with its 60.9 percent share in 1975 and even with its near 50 percent share in 1955 when Gulf oil development on a large scale was only just getting under way. Meanwhile, oil consumption within the region has grown continuously and is now at an all-time high of over 20 percent of its production, while an even higher percentage (over 25 percent) of the additional output between 1990 and 2005 has been required for local use.

In these circumstances, the near doubling of the Gulf's proven oil reserves from 1975 levels (to 100 billion tons at the beginning of 2005), is a relatively unimportant phenomenon. Most of these higher volumes of reserves are quite simply, given a current reserves-to-production ratio in the region of 82 years, required neither now, nor for the next 20 years, unless there are radical negative changes in the region's reserves situation in the short term. Thus, while the 61.4 percent share of the Gulf region in global oil reserves may be an interesting fact, it is, nevertheless, not an important one for evaluating the near-to-medium future oil development prospects in the Gulf. It may impress international oil market observers, but mainly because it clearly enhances the generally perceived need to find and re-evaluate oil reserves elsewhere, so as to continue to maintain the rest of the world's R/P ratio of approximately 23 years in the context of an annual oil production in the rest of the world that is now 2.35 times greater than that of the Gulf region. This will remain the situation in the short and medium term. For the long term, however, other issues pertaining to oil's future prospects arise and these will be discussed in the second half of this chapter.

The region's oil developments can thus be perceived by the international community as potentially important in the long term. However, this is only on the assumptions that oil demand will continue to grow strongly over the long term and that non-conventional oil production does not become economically viable on a large scale in much more extensive areas outside Canada and Venezuela.

Meanwhile, for at least the coming decade and conceivably well beyond that, the Middle East will be seen by the energy policy makers of large-scale oil-importing countries as the supply region of the last resort. This will remain the case unless, in a changed context, the upward pressures on the international oil market – created partly by the trade restraint emerging from the OPEC quota system and much more emphatically, due to the vagaries of the international trading system – are eliminated and replaced by prices based on long-run supply–price considerations, either in a fully competitive or a fully-ordered system.[17]

However, neither of these rational alternatives for the future of oil are perceived as very likely – or even very desirable – given governmental, corporate and public preferences: first, for less oil-intensive economies (oil currently supplies only 36.4 percent of total global energy supply, compared with 53 percent in 1975) and second, for oil from non-Middle East sources.

The latter development is reflected in a continuing and accelerating intra-regional trade in oil that is independent of the Gulf region. As shown in Table 9.1, this has expanded from only 402 million tons in 1975, to 936 million tons in 2005. In 1975, such trade amounted to only 26.7 percent of total international trade in oil and to only 43.7 percent of the Gulf's export trade. In 2005, the comparable percentages were 38 percent and 95.3 percent respectively.

In short, over the past 50 years, Gulf oil has enjoyed only relatively short periods of acceptability – even in the context of a rapidly rising global demand for energy in general, and for oil in particular. Already by the late 1950s, this previously acceptable role for the Middle East was already being questioned – under the impact of changing power relationships within the oil industry itself and their impact on broader geo-political and geo-economic issues. This evolving situation became even more fraught in the context of repeated supply interruptions and with the perception of Gulf producers' use of oil as a weapon in support of their political and strategic objectives.[18]

The next deterioration came when low-cost Arabian Gulf oil came to be priced in 1979–80 at levels that bore no relation to costs and which not only made alternative oil supplies, but also alternative sources of energy (notably gas), relatively much more attractive, thus enhancing the competition for Gulf oil.

As already shown in this chapter, the Arabian Gulf has still not recovered from external countries' perception – and the reality – of its lack of attraction as a supply source. Indeed, had oil production of the Soviet Union/FSU not declined from year to year from 1987 to 1996 (from 650 to 350 million tons), then competition for markets by the Arabian Gulf suppliers would have been even tougher than it was – particularly after the end of the Cold War when previous western perceptions of supply insecurity and other dangers of reliance on Soviet oil were largely eliminated. Thus, the Gulf region, which is still replete with huge reserves and despite the possibility of its oil being a nondepleting resource,[19] has basically failed to establish an international perception of itself as a reliable supplier on a continuing basis.

Nevertheless, this central element in the *realpolitik* of world oil is tempered by a widely-held conventional view of future global supply–demand relationships, viz. the inevitability of a world which cannot get by without a dramatic re-expansion of Gulf oil production and exports.[20] Moreover, the perceived international fears regarding dependence on Arabian Gulf oil are complemented by other fears stemming from the hypothesized absence of any alternative to such dependence. As shown earlier, such fears are justified by the experiences of the past 25 years and by the continuing absence of evidence to indicate any near-future radical changes in the fundamentals of the situation. Not even the re-entry of the international oil companies to the arena would instantly alter the perception of the uncertainty of supplies on a continuing basis at prices related to long-run supply prices.

However, the fears arising from the hypothesized inevitability of dependence on Gulf oil are not so soundly based. They are partly based on perceived prospects for oil industry developments, which are very generous on the demand side, compared with the average annual rate of increase in oil use of only 0.85 percent between 1975 and 2005. They also emerge in part, from underestimates of the supply side potential from outside the Arabian Gulf. What is missing in this interpretation, is a lack of consideration for the recent rapid technological developments in the exploration for, and the exploitation of, oil reserves. Such expertise, having been initiated in a few limited areas (both onshore and offshore United States, offshore Brazil and Angola, and in the North Sea), is now diffusing around most of the remaining extensive parts of the world with oil potential – except, ironically, in the Arabian Gulf, where the upstream industry has generally been starved of investments and expertise. Under such conditions, all forecasts of the inevitability of the world's future dependence on Gulf oil have a low probability of being correct: a hypothesis that is based on the wider range of considerations set out briefly in the following section of this chapter.[21]

The Long-Term Prospective Global Energy Outlook

The Global Demand for Energy

As indicated in Figure 9.2, since 1973 the rate of increase in the global use of energy has fallen well under its long-term trend of 2.2 percent a year during the period 1860–1945. This decline is in marked contrast to the ±5 percent per year growth in demand from 1945 to 1973. The probability of a return to the latter much higher annual rate of growth is now close to zero, given that it reflected a temporary set of conditions, which cannot recur. Thus, a realistic consideration of all long-term energy supply requirements – including the large oil resources of the Arabian Gulf – now has to be orientated to a modest rate of growth in the use of energy. This is required even before taking into account the possible impact of environmental concerns on non-renewable energy use together with the slowly accelerating pace of growth in the use of renewable energies, stemming from the rapidly evolving direct and indirect solar energy production technologies supported by government subsidies.

Figure 9.2
Trends in the Evolution of World Energy Use (1860–2000)

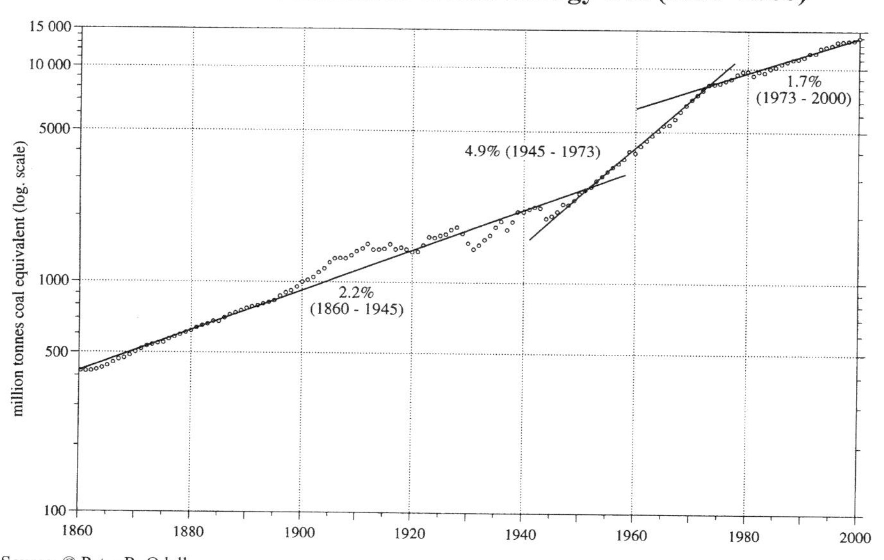

Source: © Peter R. Odell.

Therefore, the long-term future availability of oil and other carbon energy sources needs to be considered only in relation to a growth of 2 percent a year as the highest likely requirement. Moreover, the 2000 base year use of energy from which we can now evaluate the demand for coal, oil and gas in the 21st century is considerably lower than that indicated by earlier conventional forecasts. Thus, only about 26 billion barrels of oil were used in 2000, compared with the 1972 forecast of almost 140 billion. Likewise, the cumulative use of oil from 1971 to 2000 was below 700 billion barrels, rather than the anticipated 1,750 billion, so that some of the oil already discovered by 1971 still remains to be used in the 21st century. The situation is similar for coal. A 1980 World Coal Study forecast that annual use would increase from the then current level of 3000 million tons to 10,000 million by 2000. Instead, there was a mere 10 percent growth to 3300 million tons. Meanwhile, proven reserves of coal have increased by 50 percent. It is hardly surprising that views on the future availability of carbon energy and the supply potential associated with them can now be so much more relaxed than they were 30 years ago.

Global Energy Production

Figure 9.3 sets out the predicted trend in the annual production of energy supplies by source during the 21st century. This trend is based on an energy demand growth that is sustained at ±2percent a year until mid-century, and thereafter, in response to the cessation of population growth and an increasingly efficient use of energy, at a steadily declining rate of growth per decade through to 2100. Non-renewable energy sources remain overwhelmingly dominant until 2080. Alternatives to these carbon fuels will hardly exceed 25 percent of total energy supply until some time in the 2070s and, even by the end of the century, they will still account for only 40 percent of supply. Even cumulatively over the century, renewable energy sources will, as shown in Figure 9.3, supply about 30 percent of the total energy used, and over 60 percent of this will be supplied in the last three decades. Thus, unless and until the governments and peoples of the world accept the desirability of a much faster switch to renewable energy, and also take the necessary steps to implement the change, global energy use in the 21st century will remain heavily oriented towards a combination of coal, oil and natural gas. To date, there are no serious signs of these required actions for change being met.

Figure 9.3
Trends in the Evolution of Energy Supplies by Source in the 21st Century

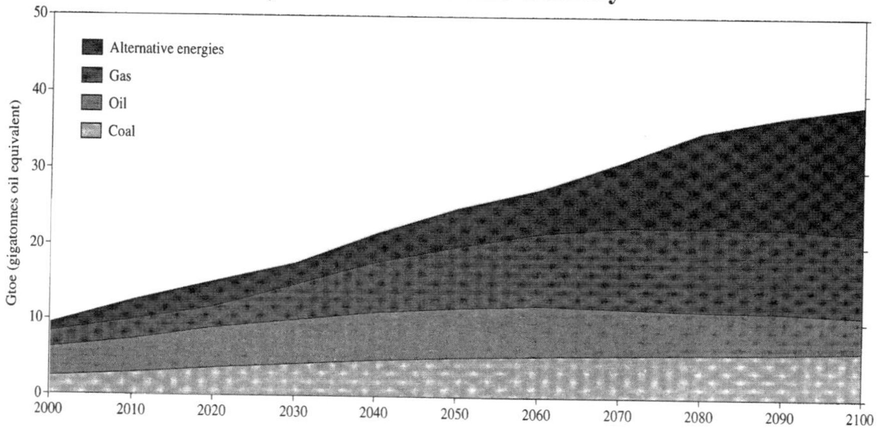

Source: © Peter R. Odell.

Figure 9.4
Cumulative Supplies of Energy by Source in the 21st Century

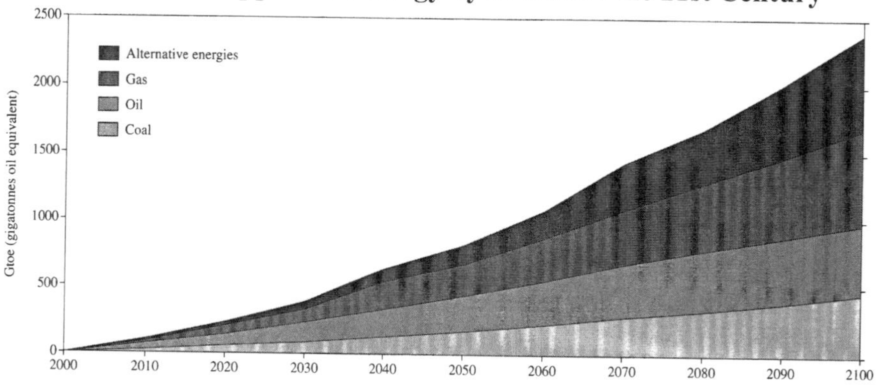

Source: © Peter R. Odell.

A Slow Expansion of Coal for Ever

Nevertheless, for both economic and environmental reasons, the relative importance of the three carbon fuels will show marked changes from the contemporary situation. Moreover, the current widely accepted conventional wisdom of inevitably constrained carbon energy supplies can be shown to be based on misconceptions concerning potentially available resources. Analysts usually indicate coal resources as an order of magnitude greater than those of oil or gas and thus implicitly, and sometimes even explicitly, assume that coal will be more important than oil and gas and eventually, become the dominant component in the 21st century carbon fuel supplies.[22] Given the general lack of acceptability of coal and its highly geographically concentrated pattern of production (over 80 percent from only ten countries), this prospect was never a realistic one. Now, even that small likelihood has been further diminished by a combination of local, regional and global environmental concerns over coal production and use. Thus, the share of coal in global carbon fuel production seems likely to increase only slowly over the century during which it will contribute only a little over 28 percent of cumulative carbon energy use.

The Role of Hydrocarbons

Thus, oil and gas between them will continue to provide most of the world's energy supply—virtually until the end of the 21st century. Consequently, during the course of the century they will, in combination, show a more-than-three-fold overall increase in their annual supply. However, the relative contributions of oil and gas to the total hydrocarbon supply will change radically, as shown in Figure 9.5. Part of this change may arise from possible long-term constraints on oil supplies, but in greater part, it is a reflection of the inherent advantages of natural gas in respect of both supply and use considerations. Supplies of natural gas will indeed, continue to expand until 2090, when its output will be some five-and-a-half times its level in 2000. The output of oil, on the other hand, seems likely to decline slowly from the 2050s, and its contribution to the annual hydrocarbon supply will ultimately fall below 30 percent. Over the century as a whole, the share of oil in the world's cumulative use of hydrocarbons will be only 32 percent.

Figure 9.5
Oil and Gas Supplies in the 21st Century

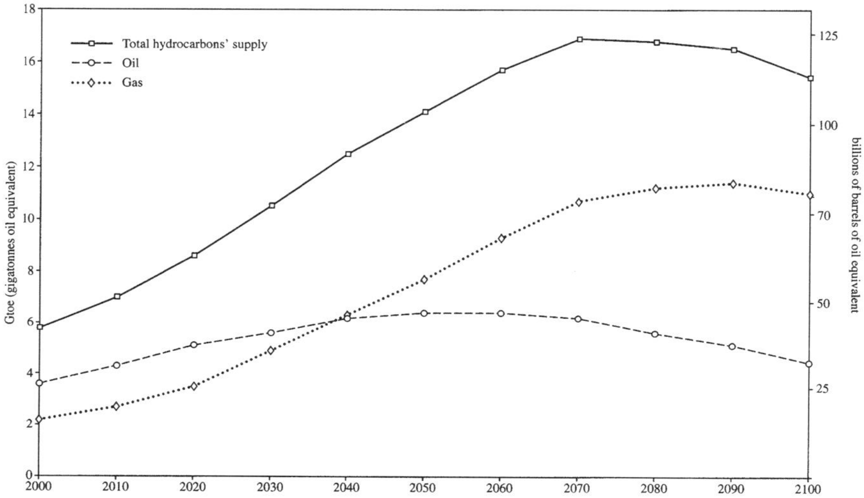

Source: © Peter R. Odell

The Declining Relative Importance of Oil

The long-term future for oil will thus be squeezed – to a greater or lesser extent, depending on environmental and economic considerations – between the undoubted solidity of the rising annual coal supply over the whole of the 21st century from the present 2.8 Gtoe to 5.4 gigatons in 2100; and from the powerful concurrent dynamics of the global gas industry with an expected increase in its annual supply from 2.4 to almost 12 Gtoe. The future of oil will thus essentially be limited by demand considerations – including geopolitical ones, as discussed in above – so that potential resource (supply-side) limitations present only a secondary issue. This is demonstrated in Figures 9.6a and 9.6b, in which the production curves cover the almost complete 200-year history of the oil industry from 1940 to 2140.

Figure 9.6a
Production of Conventional and Non-Conventional Oil (1940–2140)

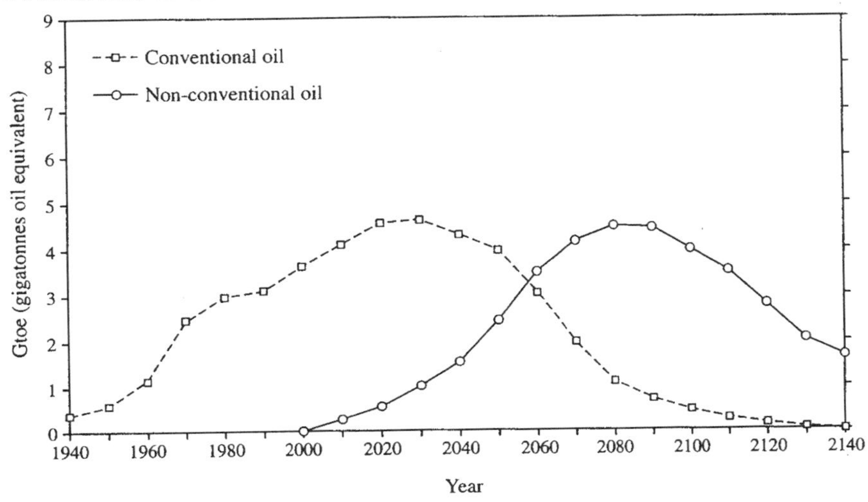

Figure 9.6b
The Complementarity of Conventional and Non-Conventional Oil Production: Giving a Higher and Later Peak to Global Oil Supplies

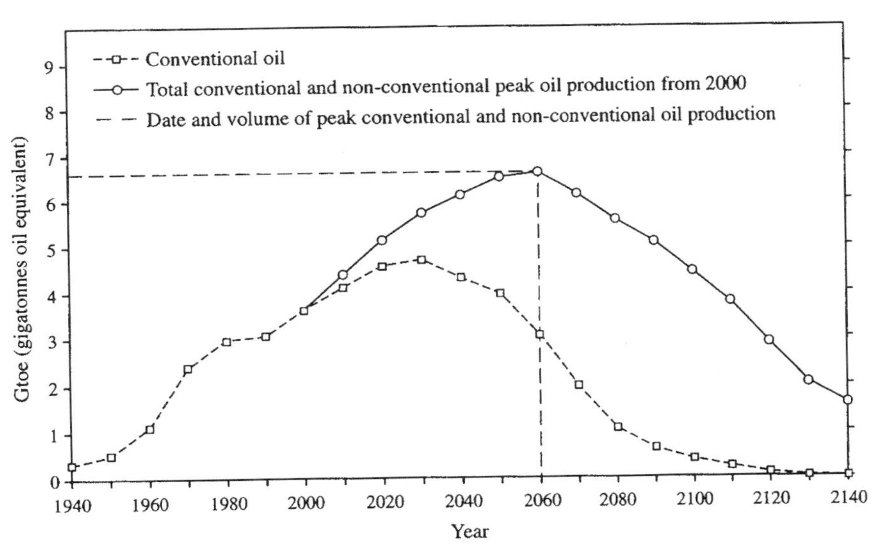

Source: © Peter R. Odell.

Conventional oil is already well on into its life cycle, but it still has around 30 years to reach its peak. By contrast, non-conventional oil production has barely started, though output is now accelerating, especially in Canada and Venezuela. With an assumption that only 3,000 billion barrels of non-conventional reserves will be recoverable (from a resource base many times larger), it will be 80–90 years before it reaches its potential peak production. This seems likely to be a little lower than the peak which will be reached by conventional oil in 2030.

Nevertheless, the two contrasting habitats of oil are essentially complementary in respect of satisfying market demand, given that customers are interested only in the personal utility of the products they need and are wholly or largely indifferent as to how and from where the products they use have been derived. Non-conventional oil will, at first, modestly supplement so-called conventional production, but post-2050 it will become the larger component of total oil supply. In general, the nearly 100-year period required for the full change from the one to the other can be portrayed as reflecting a slow, but continuing process based on the importance of both economic-geographical considerations and technological developments.

However, the oil component of energy production in the 21st century is by no means a small or short-lived one. The supply increase indicated for the first half of the century is at a rate made possible by a combination of already known reserves, appreciation of reserves and new discoveries of conventional oil, plus the steadily rising flows of non-conventional oil.

Nevertheless, the peak of oil production at around 2060 can be interpreted as being both retarded and lower than it might otherwise have been, as a result of competition from other energy sources. The rate of decline in oil production after its peak in 2060 is quite slow, so that even by 2100, the oil industry is predicated to be larger than it was in 2000. By 2100, however, in the context of resource limitations and the strong competition for markets, oil will then contribute rather less than coal to the global energy supply. Even more significantly, oil will, as shown in Figure 9.5, already have been less important than natural gas for almost 60 years. Consequently, though the present geopolitical importance of oil will undoubtedly continue in the first two decades of the 21st century, thereafter, its political significance will rapidly fade, as its contribution to the total energy supply progressively declines.

Natural Gas: The Fuel of the 21st Century

Natural gas will likely overtake coal as the second most important global energy source no later than the mid-2020s. This development reflects the near-quadrupling of worldwide proven gas reserves and the rapid expansion of European and other markets for gas between 1975 and 2005. The industry now has a generous reserves-to-production ratio of 65 years. Nevertheless, it is still more likely than not that the expansion of production in the early decades of the 21st century will continue to be limited by demand, rather than by resources. Indeed, reserves already discovered could in themselves enable global gas production to grow at 3 percent or more a year until 2025, demand permitting. Meanwhile, the continuation of additional discoveries is a certain prospect. This is a function of the geographically broadening base of gas exploration activities and the continuing opportunities for the more intensive exploitation of existing gas-rich provinces, including some hitherto thought to be fully mature, including the Gulf of Mexico, Canada, Indonesia and the North Sea, and others which have only recently been exploited on a large scale, as for example, in Iran, Saudi Arabia, the UAE, Qatar, Egypt, Malaysia and China. The mid-point of the range of estimated additional natural gas reserves indicates a volume that is about one-and-a-half times that of currently declared proven reserves. Thus, proven reserves plus only half of the estimated additional reserves are sufficient to support a conventional gas production curve which grows until at least 2050.

Figure 9.7a
Production of Conventional and Non-Conventional Gas (1940–2140)

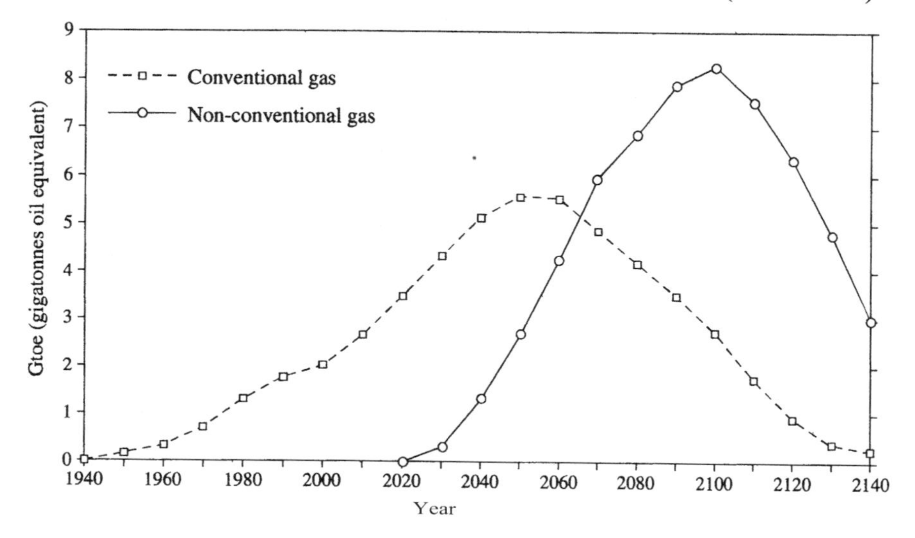

Figure 9.7b
The Complementarity of Conventional and Non-Coventional Gas Production: Giving a Higher and Later Peak to Global Gas Supplies

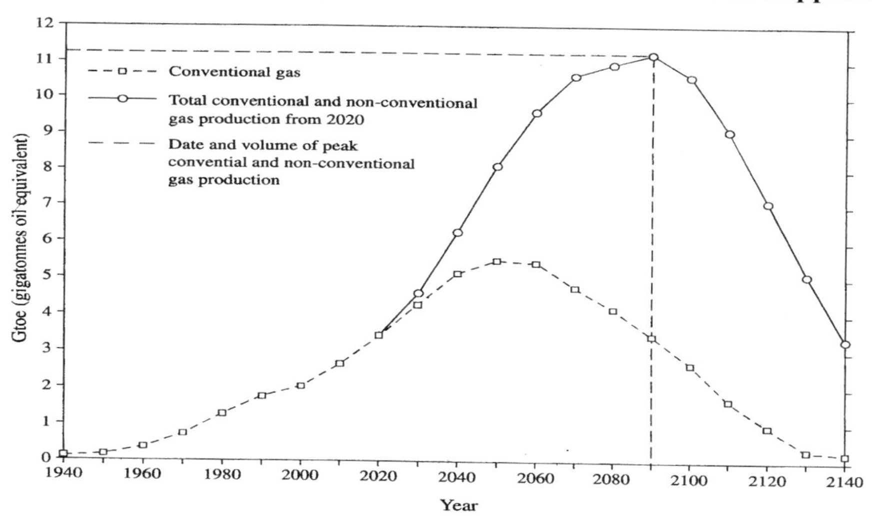

Source: © Peter R. Odell.

Figures 9.7a and 9.7b show how the more-than-200-year curve of conventional gas supply from 1940 to 2150 will be complemented by an additional production curve for non-conventional natural gas. This latter curve emerges from a conservative assumption of ultimately recoverable reserves of 650 Gtoe from coal-bed methane, tight formation gas, gas from fractured shales and ultra-deep gas. It does, however, exclude the possibility of any recovery from gas hydrates, the volumes of which have been estimated to be between 2 and 20 times larger than all other non-conventional gas resources taken together. Large scale non-conventional gas production is indicated to start in the 2020s and to become the more important component in the total gas supply before 2070. Its production will continue to grow to the end of the 21st century. In 2100 natural gas overall is predicated to supply about 50 percent of the world's carbon energy production (see Figure 9.3) and over the 21st century as a whole, about 40 percent of the cumulative total (see Figure 9.4).

Thus, gas will undoubtedly be the fuel of the 21st century (as coal was of the 19th century and oil of the 20th century), despite the demand-limiting assumptions on the exploitation of the world's gas resources, as specified above. The Gulf region, with almost 40 percent of the world's current proven reserves of conventional gas, should be capable of greatly enhancing its current contribution (of just 10 percent) to global gas supply.

Indeed, such increased supplies of natural gas from the Gulf could be critical in ensuring the ability of global production to grow sufficiently quickly through to mid-century, when conventional gas output is likely to peak at about three times its present level of 2.4 Gtoe (see Figure 9.7a). Thereafter, however, in order to sustain the predicated 2 percent per annum expansion in supplies of carbon fuels, there is likely to be a need for additional gas production from the initial exploitation of gas hydrates. A further 50 years of continuing scientific advances and engineering capabilities is likely by then to ensure a capability for achieving the recovery of at least a small part of that massive potential source of energy, in the form of the world's preferred carbon fuel, for both commercial and environmental reasons. It thus seems more likely than not to be brought to the market—assuming, of course, that natural gas can then still compete with the alternative, renewable energy sources.

Security of Demand: Where does the Gulf Region Stand in 2006?

- Recent international political events as well as extraordinary developments in the oil market continue to accentuate long-held western fears for the security of oil supplies from the Arabian Gulf region. Policy makers in many countries are already actively seeking to minimize dependence on oil, in general, and on Gulf oil, in particular.
- Present and potential supplies of both conventional and non-conventional oil in non-Gulf locations could thus be preferentially exploited, despite their much higher costs of production and/or transportation. Additionally, greater efforts are now being made to stimulate the rate of substitution of oil by natural gas, the availability of which is geographically more dispersed than that of oil and the use of which is more environmentally friendly. In this context, the Arabian Gulf's gas reserves need to be more effectively and extensively exploited through pipelines and maritime links with European and Asian markets, thus supplementing revenues/profits from oil exports.
- The high oil prices over the past two years have already accelerated oil and gas developments elsewhere for both economic and political reasons – most notably, but not exclusively – in Russia and other former Soviet republics. Moreover, such high prices are also constraining oil demand (as in the period 1979–85), partly by developments in the technologies of oil use and partly by the substitution of alternative coal and renewable energy resources.

- Gulf interests in the long-term exploitation of its large oil resources are in the process of being undermined by the recent dramatic expansion of the hitherto little known and relatively little used electronic crude oil market exchanges (particularly the NYMEX and the IPE). Trading in "paper" barrels on these exchanges, principally by "non-commercial" participants, is fundamentally undermining producer/consumer relationships and is making oil pricing almost entirely a speculative phenomenon. Gulf oil producers need to restrain this adverse external development for their long-term interests, primarily by returning exclusively to direct relationships with "physical" oil purchasers. This is necessary in order to avoid losing most of the economic rent to the speculators (estimated to be at least $25 billion in 2005) and to stabilize prices at levels mutually acceptable to exporters and importers, thus helping to secure the expansion of the world's oil markets on a long-term basis.

Such an approach to the pricing and marketing of crude oil should become increasingly likely as international oil trade becomes state-oriented, rather than private sector-oriented. Concurrently, national oil corporations in the Gulf producing countries should also increasingly accept joint venture investments from the countries of the developing world – most notably China and India – in which most incremental growth in global oil demand will occur over the coming decades.

CONTRIBUTORS

H.E. SHEIKH AHMAD FAHAD AL AHMAD AL SABAH served as the President of the OPEC Conference and Secretary General of the Organization of Petroleum Exporting Countries during 2005. He was the Minister of Energy of the State of Kuwait and Chairman of the Kuwait Petroleum Corporation from 2003 to 2006. Prior to this, he was the Minister of Information and Chairman of the National Council for Culture, Arts and Letters from 2001 to 2002. In addition, he held the position of Acting Minister of Oil during the year 2002.

Sheikh Ahmad Fahad Al Ahmad Al Sabah was appointed Chairman of the Public Authority for Youth and Sports in 2000, after serving for eight years as Vice Chairman of the Authority from 1992. In 1985, he became an officer in the Kuwaiti Army after graduating in Political Science from Kuwait University in the same year.

HERMAN T. FRANSSEN is the President of International Energy Associates Inc. and Director of Petroleum Economics Ltd (PEL) in London. He is a Senior Fellow with Center for Strategic and International Studies (CSIS); an Adjunct Scholar with the Middle East Institute in Washington, DC; an Associate of GDP Associates in New York; a Senior Associate with Middle East Consultants (MEC); and a Visiting Fellow with the Centre for Global Energy Studies (CGES) in London. Dr. Franssen was Senior Economic Advisor to the Minister of Petroleum and Minerals of the Sultanate of Oman (1985–1996) and originated the formation of the Independent Petroleum Exporting Countries (IPEC), an informal grouping of 14 non-OPEC oil-exporting countries and provinces.

Dr. Franssen was the International Energy Agency's Chief Economist (1980–1985) and was responsible for the first major IEA *World Energy Outlook* in 1983. He served as Director of the Office of International Market Analysis at the US Department of Energy (1978–1980). He was a Research Associate on Energy and Science Policy at the US Congressional Research Service. As Specialist in Energy and Environmental Policy, he acted as an energy advisor to both the Senate and House committees of the US Congress. Dr. Franssen hails from the Netherlands, where he attended the University of Amsterdam. He received a BA from Macalester College in St. Paul, Minnesota, and an MA, MALD and Ph.D from the Fletcher School of Law and Diplomacy in Medford, Massachusetts.

TARIK M. YOUSEF is Dean of the Dubai School of Government. He is on leave from Georgetown University where he is Shaykh Al-Sabah Chair of Arab Studies in the Center for Contemporary Arab Studies. He received his Ph.D in Economics from Harvard University in 1997 and specializes in development economics and economic history with a particular focus on the Middle East.

Dr. Yousef has published many book chapters and articles in journals, including the *American Political Science Review*, *Journal of Economic Perspectives* and the *Journal of Money, Credit and Banking*. He completed a World Bank Flagship report for the 2003 Annual Meetings of the IMF and World Bank in Dubai: *Unlocking the Employment Potential in the Middle East and North Africa: Toward A New Social Contract* (World Bank, 2004). At present, he is an advisor on Middle Eastern economic affairs to governments and international organizations.

EDWARD L. MORSE is Managing Director and Chief Energy Economist at the New York-based investment banking firm, Lehman Brothers. He has had an extensive career in the energy sector, spanning academia, government, publishing and business. He joined Lehman Brothers in 2006 from Hess Energy Trading Company (HETCO), where he had been Executive Advisor. Dr. Morse received his Ph.D. in International Affairs at Princeton University, where he later taught the politics of international economic relations. He served on the senior research staff of the Council on Foreign Relations in New York, after which he joined the Department of State where he served in various capacities including as Deputy Assistant Secretary for International Energy Policy. In this position he represented the United States at the International Energy Agency and in various bilateral energy working groups.

After leaving the US government, Dr. Morse joined Phillips Petroleum Company, and later co-founded PFC Energy, a consulting firm based in Washington, DC. PFC Energy later purchased *Petroleum Intelligence Weekly*, where he served as President, CEO and Publisher before joining HETCO. The author or co-author of six books and dozens of scholarly articles, he also chaired a joint task force of the Council on Foreign Relations and the James A. Baker III Institute at Rice University on US energy policy. He currently chairs a new task force on the same subject at Rice University.

PEDRO ANTONIO MERINO GARCÍA graduated from the Universidad Autónoma de Madrid as Licenciado en Ciencias Económicas and obtained his Master's degree in Business Administration from the Instituto Universitario de Administración de Empresas (I.A.D.E.). Mr. Merino was appointed Professor of Economics at the University of Extremadura from 1986–1987. In 1988 he graduated from the Spanish Civil Service as an Economista del Estado (Senior Economist) and was posted at the Ministry of Economy and Finance (Spanish Treasury) between 1988 and 1992. For the next two years, he served as Advisor to the Executive Director of the International Monetary Fund. In 1995, he became Deputy Director of Coordination and Studies at the Spanish Treasury, Alternate Director at the Board of the European Investment Bank, and Alternate Member of the Monetary Committee of the European Union. Two years later, Mr. Merino was appointed as Secretary of the Interministerial Commission for the Introduction of the Euro and Deputy Director for Coordination with Monetary Organizations at the Spanish Treasury. In 1999, he became the Director of the Economic Studies Department at the Spanish energy company Repsol YPF.

Among Mr. Merino's recent publications are: "Some Comments on Global Energy Forecast for the Period 2000–2020," Boletín de Información Comercial Española Ministerio de Economía. (2001); "Recent Evolution and Perspectives of the Oil Market (2000–2001)," Boletín Económico de Información Comercial Española. Ministerio de Economía (2001); "Missing Barrels and Statistical Deficiencies in the Oil Market," Fuentes

Estadísticas (2002); "Evolution and Forecasts of the Energy Sector Growth in the People's Republic of China," Cuadernos de Energía, Club Español de la Energía (2004); "Analysis of the Projections of Non-OPEC Petroleum Production, Short and Long Term," Boletín Económico de Información Comercial Española, Ministerio de Economía (2005); and "Explaining the So-called "Price Premium" in Oil Markets," *OPEC Review: Energy Economics and Related Issues* (2005).

JEAN-PIERRE FAVENNEC began his career as a consultant in the oil industry and has worked on several projects in gas production, refining and petrochemical complexes, and conducted studies on the strategies, pricing and profitability of these sectors. He has worked as a project manager in over fifty different countries in South America, Africa, the Middle East, Asia and Europe. Since 1990, Jean-Pierre Favennec has been a professor at the Institut Français du Pétrole (IFP) School where he worked initially as Head of Economics and Corporate Management. He was responsible for conducting training seminars for major energy companies in France and abroad. In October 2000, he was appointed Director of the Centre for Economics and Management at the IFP School. He has participated in many international conferences, speaking about the energy, oil and gas industries.

Professor Favennec's expertise includes energy economics, geopolitics and strategic aspects of energy. In particular, he has worked on the downstream sector of the hydrocarbons industry, and written papers on related subjects. He has edited two books: *Refining Economics and Management and Research* and *Production of Oil and Gas: Reserves, Costs, Contracts*. More recent books include *Energy Markets*, published in January 2006 (French version by Edition Technip, English version forthcoming). A new book *Energy Geopolitics* is due to be published in December 2006 (English version to follow). Professor Favennec holds a Master's degree in Chemical Engineering from Ecole Nationale Supérieure des Industries Chimiques, in Nancy, France and a degree in Oil Economics from the IFP School.

NAWAF E. OBAID is a Saudi national security and intelligence consultant based in Riyadh, Saudi Arabia. He is currently the Managing Director of the Saudi National Security Assessment Project. He is also an Adjunct Fellow in the Arleigh A. Burke Chair in Strategy at the Center for Strategic and International Studies (CSIS) in Washington, DC. He was a former Research Fellow at the Washington Institute for Near East Policy (WINEP). He also served as the Project Director for a major study "Sino-Saudi Energy Rapprochement and the Implications for US National Security," conducted for the Advisor and Director for Net Assessments to the US Secretary of Defense. Many of his studies, reports and opinion pieces have appeared in newspapers such as *The Washington Post, New York Times, The Financial Times* and the *International Herald Tribune*.

He has authored a book on Saudi oil policy entitled: *The Oil Kingdom at 100: Petroleum Policymaking in Saudi Arabia* (WINEP Editions). He has recently published a book entitled *National Security in Saudi Arabia: Threats, Responses and Challenges* co-authored with Dr. Anthony Cordesman (Praeger and CSIS Publications, 2005). He is working on another book entitled *The Struggle for the Saudi Soul: Royalty, Militancy and Reform in the Kingdom.* He holds a BS degree from Georgetown University's School of Foreign Service, an MA from Harvard University's Kennedy School of Government and has completed doctoral courses at the Security Studies Program of the Massachusetts Institute of Technology (MIT).

ALOULOU FAWZI is an Energy Economist at the International, Economic and Greenhouse Gases Division of the Office of Integrated Analysis and Forecasting (OIAF) of the Energy Information Administration (EIA). Responsible for co-authoring the "World Natural Gas and Oil Markets Outlook" chapters of the EIA annual publication the *International Energy Outlook* (IEO), he works with the team of modelers who develop the EIA long-term world energy forecast. Mr. Fawzi is also the EIA's China expert, monitoring the country's energy resources, consumption patterns, trade, technology's use and investment strategies. Since 2004, he has served as

Advisor to Tsinghua–BP Clean Energy and Education Center at Tsinghua University in Beijing.

Prior to joining the EIA in 2001, Mr. Fawzi was a Research Associate at Cambridge Energy Research Associates (CERA) in Massachusetts, where he monitored market fundamentals, as well as political and policy developments to assess supply, demand and the price of crude and refined products. While at CERA he authored reports on the corporate strategies of national oil companies, Japan's activities in the Middle Eastern oil and gas sector, and the taxation of petroleum products. He also served as Energy Advisor at the Prime Minister's Department in Malaysia and reviewed Malaysia's investment programs and energy policies (Vision 2020). Mr. Fawzi holds a Bachelor of Arts degree from the University of California at Berkeley and a Masters in Public Administration from the Kennedy School of Government at Harvard University.

KANG WU has been a researcher at the East-West Center since 1991 and is currently a Senior Fellow. He conducts energy and economic research on the Asia-Pacific region and covers energy policies, security, demand, supply, trade and market developments, energy–economic links, oil and gas issues, and the environmental impact of fossil energy consumption. An energy expert on China, Dr. Wu supervises the China Energy Project at the Center and also deals with energy sector issues in other Asia-Pacific countries. As an energy economist, his work includes energy modeling and Asia-Pacific energy demand forecasting. Dr. Wu received his Ph.D. and MA degrees in Economics from the University of Hawaii at Manoa and his BA in International Economics from Peking University, China.

Dr. Wu is the author and co-author of several journal articles, project reports, professional studies, conference papers, books and other publications. He is a frequent speaker at international conferences, forums, workshops and training programs. His research work has been cited by the press and industrial media, including *Time* magazine, *New York Times*, *The Wall Street Journal*, *The Asian Wall Street Journal*, *International Herald Tribune*, *Far Eastern Economic Review*, *Journal of Commerce*, *The Strait Times*, *Honolulu Advertiser* and the *Pacific Business News*. He has also been quoted by news agencies and broadcast channels such as Radio Free Asia, Reuters, Voice of America, BBC, Dow Jones Energy Services, Bloomberg News, KGMB 9 News, and KITV 4/ABC News.

JIT YANG LIM is a Senior Consultant at FACTS Global Energy (FGE), based in Singapore. Prior to joining FGE, Dr. Lim was the Manager of a research unit at Goodyear Tire Company, where he was responsible for the modeling and forecasting of commodity prices at its global purchasing office. Previous industrial positions include that of a Statistical Engineer at Advanced Micro Devices (AMD) where he conducted process improvement projects on various different issues, as well as provided statistical consultancy services and problem-solving skills to employees, as part of the Total Quality Management initiatives in the company. Dr. Lim started his professional career in the IT industry, where he was an Analyst Programmer who carried out software development in the commercial environment.

Dr. Lim holds an M.Sc degree in Operations Research, an MBA, and a Ph.D in Economics. Since joining FGE, he has worked on a number of projects, which include: Market Feasibility Study for Investment in Refining, Study of the Far East Oil Market Structure, Product Quality Study in the East of Suez, and Market Analysis Study for Refinery Expansion. His research focuses on the downstream oil industry in the Asia-Pacific, specifically in the analysis of petroleum demand, and the modeling and forecasting of oil prices and refining margins.

PETER R. ODELL is a Professor Emeritus of Erasmus University in Rotterdam where he was the Director of the University's Centre for International Energy Studies. Over the decades he has advised many public and private bodies on energy-related issues and has lectured on his research interests at more than 100 academic and professional institutions around the world. He has also been called upon to present evidence on conflicts between countries and companies on international oil and gas issues in many arbitration tribunals and the International Court of Justice. In 1991 he was honored by the International Association for Energy Economics for his "outstanding contributions to the subject and its literature" and in 1994 he was awarded the Royal Scottish Geographical Society's Centennial Medal for his studies on North Sea oil and gas.

His extensive research and prolific publications on a broad range of economic and geopolitical issues, relating to global and European energy, date back to the early 1960s and the 1970s. His best known and best-selling book was *Oil and World Power* (Penguin Books, London) which ran to eight editions and 13 translations between 1970 and 1986. More recently, in 2001–2002 he published a two-volume collection of 70 of his selected studies and commentaries, entitled *Oil and Gas: Crises and Controversies, 1961–2000* (Multi-Science Publishing, UK). In 2004, he published the book *Why Carbons Fuels Will Dominate the 21st Century's Global Energy Economy* (Multi-Science Publishing), in which he attempts to draw together the economic, political and environmental issues which will influence global energy prospects in the medium and longer term.

NOTES

Chapter 2

1. Throughout this paper, the Gulf economies refer to the member countries of the Gulf Cooperation Council (GCC): Bahrain, Kuwait, Qatar, Oman, Saudi Arabia and the UAE.

2. On the exceptional conditions of the past few years in the Middle East (including the Gulf) and their implications for short-term and long-term policies, see *World Bank, Economic Developments and Prospects: Oil Booms and Revenues Management* (Washington, DC: World Bank, 2005).

3. See Vahan Zanoyan, "After the Oil Boom: The Holiday Ends in the Gulf," *Foreign Affairs* vol. 74, no. 6 (1997): 2–7, for an earlier perspective on the structural challenges facing the Gulf, and Tarik Yousef, "Development, Growth and Policy Reform in the Middle East and North Africa since 1950," *Journal of Economic Perspectives* vol. 18, no. 3 (2004): 91–116, for both a more recent perspective and a discussion of the link between oil prices and reform.

4. Throughout this paper, the term "globalization" is used in a very broad sense to refer to the rapid integration of international commodity and factor markets through trade, migration and capital flows. This covers both formal and informal channels as well as legal and illegal transactions.

5. For an introduction to different conceptualizations of the notion of economic security, see Helen E.S. Nesadurai, "Introduction: Economic Security, Globalization and Governance," *The Pacific Review* vol. 17, no. 4 (2004): 460–482. For a clear statement on the transformation in the meaning of economic security in light of globalization, see Miles Kahler, "Economic Security in an Era of

Globalization: Definition and Provision," *The Pacific Review* vol. 17, no. 4 (2004): 486.

6. For a discussion of these development strategies and their shortcomings, see Charles F. Doran, "Economics and Security in the Gulf," in David E. Long and Christian Koch (eds) *Gulf Security in the Twenty-First Century* (Abu Dhabi: The Emirates Center for Strategic Studies and Research, 1997).

7. The lack of economic diversification (including non-oil manufacturing exports) is one interpretation of the so-called "natural resource curse," in which countries with natural resource abundance appear to underperform in long-run GDP growth. This happens because first, the positive wealth shocks from the export of natural resources raise demand for non-tradeable products, drawing skilled workers, physical capital investment and entrepreneurial ability from other sectors; and second, the high spending leads to the loss of competitiveness due to the appreciation of the real exchange rate. Thus, the tradeable sector, especially manufacturing activity, declines and economic growth suffers. See Jeffrey Sachs and Andrew M. Warner, "Natural Resource Abundance and Economic Growth," Harvard Institute for International Development Discussion Paper (1995).

8. For an analysis showing that the development in the growth of the non-oil sectors in the long run cannot be determined by the expansion of government expenditures, see Ugo Fasano and Qing Wang, "Fiscal Expenditure Policy and Non-Oil Economic Growth: Evidence from GCC Countries," *IMF Working Paper* 01/195 (Washington, DC: International Monetary Fund, 2001). On the impact of the oil price decline in the 1980s, see World Bank, *Claiming the Future: Choosing Prosperity in the Middle East and North Africa* (Washington, DC: World Bank, 1995).

9. It is tempting to exclude the constraint of a small domestic market from applying to Saudi Arabia. While the constraint may be less binding there relative to the smaller Gulf economies, there is little evidence that Saudi non-oil industries, especially those with public sector ownership, have relied on the larger size of the domestic market alone to remain profitable.

10. Regarding the importance of regional integration in expanding the size of the market, see Doran, op. cit.

11. On this and the role of the Asian crisis of 1997 in creating perceptions of economic insecurity, see Kahler, op. cit., 488–93.

12. For an elaboration of the shortcomings of global economic policy and the risks paused by short-term capital mobility, see Joseph Stiglitz, *Globalization and its Discontents* (New York, NY: W.W. Norton & Company, 2002).

13. See Kahler, op. cit., 489.

14. For example, the widely expressed apprehension towards globalization in recent public opinion polls in the Arab world is partly due to extensive inward-looking state intervention in the political economies of the region in the post-independence era. This legacy, however, was reinforced by the post-Sept. 11, 2001 environment that further isolated students, scholars and the larger citizenry from current debates on globalization.

15. In the Middle East, see for example Rex Brynen, "Economic Crisis and Post-Rentier Democratization in the Arab World: The Case of Jordan," *Canadian Journal of Political Science* vol. 25, no. 1 (1992): 69–97; and Iliya Harik and D.J. Sullivan (eds) *Privatization and Liberalization in the Middle East* (Bloomington, IN: Indiana University Press, 1992).

16. For a clear statement of how globalization creates winners and losers in labor markets and the resulting social tensions, see Dani Rodrik, *Has Globalization Gone Too Far?* (Washington, DC: Institute for International Economics, 1997).

17. This discussion and subsequent references to labor markets in the Gulf countries draw heavily on World Bank, *Unlocking the Employment Potential in the Middle East and North Africa—Toward a New Social Contract* (Washington, DC: World Bank, 2004). See also M. Girgis, F. Hadad-Zervose and A. Coulibaly, "A Strategy for Sustainable Employment for GCC Nationals," *Working Paper*, World Bank, 2003.

18. For further discussion of employment as the region's common political and security concern, and the implications of this domestic situation for the region as a whole, see Ibrahim Karawan, "The Erosion of Consensus: Perceptions of GCC States of a Changing Region," in Lawrence Potter and Gary Sick (eds) *Security in the Persian Gulf: Origins, Obstacles, and the Search for Consensus* (New York, NY: Palgrave, 2002).

19. On the pace of reforms in the Middle East in the 1990s, including an outline of pending areas that require attention by policy makers in the future, see Tarik Yousef, "Structural Reforms, the Investment Climate and Private Sector Development in the Arab World," in Augusto Lopez-Claros and Klaus Schwab (eds) *The Arab Competitiveness Report* (Switzerland: Palgrave Macmillan, 2005).

20. For a succinct description of this framework, see World Bank, *Jobs, Growth and Governance in the Middle East and North Africa—Unlocking the Potential for Prosperity* (Washington, DC: World Bank, 2003).

21. See Augusto Lopez-Claros and Klaus Schwab (eds) *The Arab Competitiveness Report* (Switzerland: Palgrave Macmillan, 2005).

22. World Bank, *Trade, Investment and Development in the Middle East and North Africa—Engaging with the World* (Washington, DC: World Bank, 2003).

23. See Ugo Fasano et al., "Monetary Union among Member Countries of the Gulf Cooperation Council," *IMF Occasional Paper* no. 221 (Washington, DC: International Monetary Fund, 2003).

24. For a recent update on progress with diversification in the Gulf countries, see Ugo Fasano and Zubair Iqbal, "GCC Countries: From Oil Dependence to Diversification" (Washington, DC: International Monetary Fund, 2003).

25. For evidence supporting the view that government expenditures in the GCC is pro-cyclical in that they follow the course of oil revenues, see Ugo Fasano and Qing Wang, "Testing the Relationship Between Government Spending and Revenue: Evidence from GCC Countries," *IMF Working Paper* 02/201 (Washington, DC: International Monetary Fund, 2002).

26. Ugo Fasano, "Review of the Experience with Oil Stabilization and Savings Funds in Selected Countries," *IMF Working Paper* 00/112 (Washington, DC: International Monetary Fund, 2000).

27. The discussion of gender issues draws heavily on World Bank, *Gender and Development in the Middle East and North Africa: Women in the Public Sphere* (Washington, DC: World Bank, 2004).

28. For a detailed analysis of past problems in building institutions and the difficulties the Gulf countries face in reforming institutions in the future, see Dirk Vandewalle, "Social Contracts, Institutional Development, and Economic Growth and Reform in Middle East Oil Exporters," Unpublished Paper, Dartmouth College (2003).

Chapter 3

1. The literature on the so-called resource curse is vast and growing. Extremely useful is Paul Stevens' study *Resource Impact – Curse or Blessing? A Literature Survey* (London: International Petroleum Industry Environmental Conservation Association, 2003), 33. (www.ipieca.org/downloads/social/PStevens_resourceimpact_final.doc). Much of the literature is on African development and sustainability, although research into so-called "Petro-states" including the members of the Organization of Petroleum Exporting Countries (OPEC) has grown with the decline of oil prices in the 1980s and 1990s and the recent rise in prices. A critical work remains A.H. Gelb and Associates, *Windfall Gains: Blessing or Curse?* (New York, NY: Oxford University Press, 1988).

2. A recent and convincing argument about the provision of global public goods by the United States to the rest of the world can be found in Michael Mandelbaum, *The Case for Goliath: How America Acts as the World's Government in the Twenty-first Century* (New York, NY: Public Affairs Books, 2006).

3. I am grateful to Nader Sultan, former Chief Executive of KPC, for this example. Clearly the comparison of these two companies is fraught with landmines. BP is a private company, once owned by the state, but organized as a private stock company, with a complex history. As a now publicly-traded company, its accounts are transparent. KPC is a state-owned corporation built on nationalized upstream assets and its accounts are opaque. Nonetheless, the overview presented is designed to highlight the adaptability of private sector companies in confronting market competition and challenges in comparison to a state-owned company responsible for the largest part of the income of its government shareholder.

4. There are many complex issues associated with this problem, including the debate within Kuwait over foreign investment, which is itself tied to the perceptions and goals of Members of the Kuwaiti Parliament vis-à-vis the country's ruling family. Even so, these lessons seem to transcend these many micro issues.

5. See Dermot Gately, "OPEC's Incentives for Faster Output Growth," *Energy Journal*, vol. 25, no. 2 (September 2004): 75–96.

6. For the full text of this speech see the website of the Ministry of Petroleum and Mineral Resources, Saudi Arabia (www.mopm.gov.sa).

7. The arguments developed in this section are derived from Amy M. Jaffe and Edward L. Morse, "OPEC in Confrontation with Globalization," in Jan H. Kalicki and David L. Goldwyn (eds) *Energy and Security* (Baltimore, MD: The Johns Hopkins University Press, 2005), 65–96. Also see Edward L. Morse, "Energy Breaks the Economic Rules," *SAISPHERE* 2005 (January 2006), 45–47.

8. The literature on the subject of OPEC's discriminatory policies is rich, long and controversial. An important statement on this subject can be found in Amy M. Jaffe and Ron Soligo, "A Note on Saudi Price Discrimination," *Energy Journal*, vol. 21, no. 1 (January 2000), 121–134. Other studies of OPEC countries' market discrimination include: Maurice A. Adelman, *The Genie Out of the Bottle: World Oil since 1970* (Cambridge, MA: MIT Press, 1995); Jacques Cremer and Djavad Salehi-Isfahani, *Models of the Oil Market* (Amsterdam: Harwood Academic Publishers, 1991); George Daley, James M. Griffin and Henry B. Steele, "Recent Oil Price Escalations: Implications for OPEC Stability," in James M. Griffin and David J. Teece (eds), *OPEC Behavior and World Oil Prices* (London: George Allen & Unwin, 1982), 145–174; James M. Griffin and Henry B. Steele, *Energy Economics and Policy*, second edition (Orlando, FL: Academic Press, 1986); Paul Horsnell and Robert Mabro, *Oil Markets*

and Prices: The Brent Market and the Formation of World Oil Prices (Oxford: Oxford University Press, 1997); Theodore Moran, "Modelling OPEC Behavior: Economic and Politic Alternatives," in James M. Griffin and David J. Teece (eds), op. cit., 954–963, and Philip K. Verleger, "The Evolution of Oil as a Commodity," in Richard L. Gordon, Henry D. Jacoby and Martin B. Zimmerman (eds), *Energy: Markets and Regulation* (Cambridge, MA: MIT Press, 1987), 161–186.

9. Among the many studies on the origins of OPEC, a convincing argument along these lines is found in Emma B. Brossard, *Petroleum: Politics and Power* (Tulsa, OK: Pennwell Books, 1983), 121–157; Also see two enduring classics: Ian Seymour, *OPEC: Instrument of Change* (New York, NY: St. Martin's Press, 1981) and Ian Skeet, *OPEC: Twenty-Five Years of Prices and Politics* (Cambridge: Cambridge University Press, 1988).

10. For discussions of OPEC in the context of the North–South Dialogue in the 1970s, see Ruth W. Arad, Uzi B. Arad, Rachel McCulloch, Jose Pinera and Ann Hollick, *Sharing Global Resources* (New York, NY: Council on Foreign Relations, 1979), and Roger Hansen (ed), *Rich and Poor Nations in the World Economy* (New York, NY: Council on Foreign Relations, 1978).

11. In the category of "strange but true," the Cold War history of the United States is replete with evidence of how the US government urged the Middle East members of OPEC, particularly Iran, to raise prices in the period 1971–75 in order to strengthen the countries in the region as a bulwark against Russian influence. Among the articles on the subject written at the time are V.H. Oppenheim "Why Oil Prices Go Up: The Past: We Pushed Them, *Foreign Policy*, no. 25 (Winter 1976–77), 24–57 and Steven D. Krasner, "A Statist Interpretation of American Oil Policy and the Middle East," *Political Science Quarterly*,

vol 94, no. 1 (Spring 1979), 77–96. Francisco Parra, a Venezuelan and former OPEC Secretary General traces various episodes during which Washington opted for higher prices. See Francisco Parra, *Oil Politics: A Modern History of Petroleum* (New York, NY: IB Tauris, 2004). See especially pages 110–136.

12. Private access to the archives of US President Ronald Reagan has provided some writers with the necessary evidence to demonstrate this point. See especially Peter Schweizer, *Victory: The Reagan Administration's Secret Strategy that Hastened the Collapse of the Soviet Union* (New York, NY: Atlantic Monthly Press, 1994), 96–111.

13. See David Rogers, "US Annual War Spending Grows," *Wall Street Journal* (March 8, 2006): 1.

14. Graham E. Fuller and Ian O. Lesser, "Persian Gulf Myths," *Foreign Affairs*, vol. 76, no. 3 (May/June 1997): 144–153. A similar conclusion can be found in Shibley Telhami, "The Persian Gulf: Understanding the American Oil Strategy," *The Brookings Review*, vol. 20, no.2 (Spring 2002): 32–35.

15. The ongoing activities of the Working Group on Governance and Transparency of the Institute for International Finance (IIF) can be found on its website (www.iif.com).

16. The recent economic boom in the Middle East has been so recent and so extraordinary a phenomenon that its recognition has been largely found in the news media rather than in scholarly research. Reports on individual countries abound, including the many reports on the robust growth of the Saudi economy found in the literature of numerous financial institutions, including especially the Saudi SAMBA financial group and reports by its chief economist, Brad Bourland. Many regional banks are also now issuing country reports on the investment

climate in GCC member states, which can serve as an import source of data and analysis. Of course publications such as *Arab News* (www.arabnews.com) also provide ongoing information about the equities and real estate markets in the region. However, there is still a long lag in data availability. At the time of writing this article, the most recent writings of the IIF that analyzed the economies of the region were more than one year old.

17. The most extensive analysis on the operation of cycles within the petroleum sector remains Richard Gordon et al., *World Petroleum Markets: A Framework for Reliable Projections*, The World Bank Technical Paper no. 92, Industry and Energy Series (Washington, DC: World Bank, 1988).

18. I have drawn heavily in this section on A.F. Alhajji, "What Have we Learnt from the Experience of Low Oil Prices? *OPEC Review* (September 2001): 193–220. While I share and agree with Alhajii's analysis, I differ with him significantly on his policy conclusions.

19. Alhajii, op. cit.,193.

20. See Mohammed Barkindo, "Energy Supply and Demand Security," EUROPIA Conference, London, February 15–16, 2006 (http://www.opec.org/opecna/Speeches/2006/Europia.htm).

21. The underlying problem is that the state's presence in the oil sector and the oil sector's position in the region's individual economies suffocate the private sector. This is the way the mechanics of the oil curse, discussed at the start of this paper continue to stifle growth. See A.F. Alhajji, "The Impact of Dualism on the GCC Private Sector?" *Oil, Gas and Energy Law Intelligence* (OGEL) vol. 3, no. 2 (June 2005).

22. A recent and excellent overview of this problem as recognized in the United States, can be found in Alan P. Larson and David M.

Marchick, *Foreign Investment and National Security: Getting the Balance Right* (New York, NY: Council on Foreign Relations, 2006).

23. A discussion of these issues can be found in the *Gulf Yearbook*, 2005–2006 (Dubai: Gulf Research Center, 2006).

24. "Mid-East IPO obsession raises over USD 6 billion in 2005," AME Info (February 25, 2006), reporting on the first Middle East IPO Summit, Dubai (http://www.ameinfo.com/78222.html).

25. Gregory Gause recognized this problem several years ago in his study *Oil Monarchies: Domestic and Security Challenges in the Arab Gulf* (New York, NY: Council on Foreign Relations, 1994).

Chapter 4

1. API gravity is a specific gravity scale developed by the American Petroleum Institute (API) for measuring the relative density of various petroleum liquids. It is conversely related to crude density (d), or specific weight (mass/volume), at 15°C (60°F) and is calculated through the following formula: $°API = \frac{141.5}{d} - 131.5$

   Therefore, the lighter the crude, the higher is the °API value.

2. Energy Intelligence Research, *The International Crude Oil Market Handbook, 2006* (Energy Intelligence, Fifth Edition, December 2005), E201 and E272.

3. *ENI World Oil & Gas Review 2006* (Rome: ENI, 2006) 121–151.

4. *ENI World Oil and Gas Review 2006* op. cit., 121–123.

5. Simmons & Company International, *Simmons Energy Monthly—Natural Gas Outlook* (Energy Industry Research, June 23, 2006) 58.

6. Michael Lewis, *Commodities Investment: From Passive to Active Strategies* (Frankfurt: Deutsche Bank, May 3, 2006).

7. Schwartz, Fuller, Billig and Bulger, *Oil Supply: Can Capacity Keep Pace with Demand?* PIRA New York Annual Seminar 2005 (New York, NY: PIRA, October 2005) and Bill Fuller 2006 projections.

8. Fatih Birol, *Global Energy Outlook 2005: Energy Security and Environmental Challenges* (Paris: IEA, 2005) 7.

9. International Energy Agency, *World Energy Outlook 2005, Middle East and North Africa Insights* (Paris: IEA, 2005) 138.

10. *ENI World Oil & Gas Review 2006* op. cit., 121–151.

11. Energy Information Administration, *International Energy Outlook 2006* (Washington, DC: USDOE, 2006), 155.

12. Energy Information Administration, *International Energy Outlook 2006* (Washington, DC: USDOE, 2006), 87.

13. International Energy Agency, *World Energy Outlook 2005, Middle East and North Africa Insights* (Paris: IEA, 2005), 90.

14. International Monetary Fund, *World Economic Outlook, Globalization and External Imbalances* (IMF, April 2005), Chapter IV, 170.

15. OPEC Secretariat paper "Oil Outlook to 2025," 10th International Energy Forum, Doha, April 22–24, 2006.

16. International Energy Agency, *World Energy Outlook 2004* (Paris: IEA, 2004) 430–431.

17. Petroleum Industry Research Associates, *Scenario Planning Service: Annual Guidebook 2006* (New York, NY: PIRA, 2006), 119–120.

18. International Energy Agency, *World Energy Outlook 2004* (IEA, 2004), 41.

19. International Monetary Fund, *World Economic Outlook, Spring 2006* (IMF, 2006).

20. Ben S. Bernanke, M. Gertler and M. Watson. "Systematic Monetary Policy and the Effects of Oil Price Shocks." *Brookings Papers on Economic Activity*, 1997, 1, 91–142.

21. Lewis Alexander, "Overview: Inflation and Monetary Policy," *Global Economic Outlook and Strategy*, Citigroup, June 2006, 4–13.

22. Bernanke et al., op. cit., and Mark A. Hooker, "Are Oil Shocks Inflationary? Asymmetric and Nonlinear Specifications versus Changes in Regime." *Finance and Economics Discussion Series* no. 65 (Washington, DC: Board of Governors of the Federal Reserve System, 1999).

23. Francisco Blanch and Sabine Schels, *Energy Strategist: A Growing Imbalance in the Global Oil Markets* (Merrill Lynch, June 28, 2006.)

24. International Energy Agency, Annual Statistical Supplement for 2004 and User's Guide (IEA, 2005) 64–66.

25. This causality is just an "econometric causality" analyzed through a Granger Causality Test.

Chapter 5

1. Other factors will be discussed in the following paragraph.

2. All production and consumption figures are taken from the *BP Statistical Review*.

3. Data complied from country sources.

4. Discussions held by the author with company representatives.

5. Data from the World Energy Council, adapted by the author.

6. Energy Information Administration (EIA) of the United States Department of Energy (USDOE).

Chapter 7

1. Energy Information Administration, *Annual Energy Outlook 2005 with Projections to 2025*, February 2005, Washington, DC (www.eia.doe.gov/oiaf/aeo/).

2. Energy Information Administration, *International Energy Outlook 2005*, July 2005, Washington, DC (www.eia.doe.gov/oiaf/ieo/index.html/).

3. These are the proven reserves as of January 1, 2005. See *Oil & Gas Journal*, vol. 102, no. 47 (December 20, 2004).

4. World Market Research Center, Country Reports, London, September 2005 (www.wmrc.com).

5. The author's personal interview with H.E. Abdulla Bin Hamad Al Attiyah, Second Deputy Premier and Minister of Energy and Industry, State of Qatar, December 2003, Washington, DC.

6. The Qatar Embassy in the United States, *Qatar: The Modern State* (Washington, DC: Qatar embassy, November 2004).

7. The author's personal interview with Khaldoon Al Mubarak, then Executive Vice President-Corporate of Dolphin Energy Limited at the US DOE-LNG Ministerial Summit, December 18, 2003, Washington, DC.

8. Aloulou Fawzi, "Gas to Liquids: A New Frontier for Natural Gas," *International Energy Outlook 2005* (Washington, DC: EIA, July 2005) 47 (www.eia.doe.gov/oiaf/ieo/index.html/).

9. Energy Information Administration, *The Global Liquefied Natural Gas Market: Status and Outlook* (Washington, DC: EIA, December 2003) (http://www.eia.doe.gov/oiaf/analysispaper/global/pdf/eia_0637.pdf).

10. Robert Ineson, "The Incoming Tide: LNG Surges into North America," Cambridge Energy Research Associates, Decision Brief, January 2004.

11. International Energy Agency (IEA), *World Energy Investment Outlook 2003* (Paris: IEA/OECD, 2003) 228 (www.iea.org)

12. The author's personal interview with Dr. Ibrahim B. Ibrahim, Chairman of Marketing and Vice Chair of the Board of Qatar RasGas Company, December 18, 2003, Washington, DC.

13. The author's personal interview with James T. Jensen, President of Jensen Associates Inc. January 7, 2004, Weston, Massachusetts (http://JAI-Energy.com).

14. The author's personal interview with Idriss Aljazairi, Algerian Ambassador to the United States, December 24, 2003, Washington, DC.

15. Ineson, op. cit.

16. Daniel Yergin and Michael Stoppard, "The Next Prize," *Foreign Affairs*, November/December 2003.

17. The author's interview with James T. Jensen, op. cit.

Chapter 8

1. The Asia-Pacific region is defined to include Australia, Brunei, China, Hong Kong, India, Indonesia, Japan, Malaysia, New Zealand, Pakistan, Philippines, Singapore, South Korea, Taiwan, Thailand, Vietnam and "Other" Asia, which refers to a collection of other smaller countries/territories in the Asia-Pacific region. Unless otherwise specified, "Asia" and the "Asia-Pacific region" may be used interchangeably to refer to the same definition above. For economic growth, see World Bank 2005 and for energy consumption, see IMF 2005.

2. Primary commercial energy consumption (PCEC) is defined to include the use of coal, oil, natural gas, nuclear power and hydroelectricity. Statistics are based on BP 2005.

3. See International Energy Agency, *Oil Market Report* (Paris: IEA/OECD, October, 2005); Energy Information Administration (EIA), *Short-Term Energy Outlook* (Washington, DC: USDOE, October, 2005); FACTS Inc. *Asia-Pacific Databook 1: Supply, Demand & Prices* (Honolulu, HI: FACTS, Fall 2005).

4. Kang Wu and Fereidun Fesharaki, "As Oil Demand Surges, China Adds and Expands Refineries," *Oil and Gas Journal* vol. 103, no. 28 (July 25, 2005): 20–24.

5. FACTS Inc. *Asia-Pacific Databook 1: Supply, Demand & Prices* (Honolulu, HI: FACTS, Fall 2005).

Chapter 9

1. S. H. Longrigg, *Oil in the Middle East*, 3rd Edition (Oxford: Oxford University Press, 1968).

2. Issawi, C. and M. Yeganeh. *The Economics of Middle Eastern Oil* (New York, NY: Praeger Publishers, 1962).

3. G. Lenczowski, *Oil and State in the Middle East* (Ithaca, NY: Cornell University Press, 1960).

4. E.T. Penrose, *The Large International Firm in Developing Countries* (London: Allen and Unwin, 1968).

5. Penrose, op. cit.

6. G. W. Stocking, *Middle East Oil: A Study in Political and Economic Controversy* (London: Allen Lane, 1971).

7. B. Shwadran, *The Middle East, Oil and the Great Powers, 3rd Edition* (New York, NY: Wiley, 1979).
8. H. Lubell, *Middle East Oil Crises and W. Europe's Energy Supplies* (New York, NY: Elsevier Press, 1971).
9. A.H. Cordesman, *Energy Developments in the Middle East* (Westport, CT: Praeger Publishers, 2004).
10. Penrose, op. cit.
11. M. A. Adelman, *The Genie Out of the Bottle: World Oil since 1970* (Cambridge, MA: MIT Press, 1995).
12. M. A. Adelman, *The World Petroleum Market* (Baltimore, MA: Johns Hopkins University Press, 1972). See also Adelman, *The Economics of Petroleum Supply* (Cambridge, MA: MIT Press, 1993).
13. E. Kanovsky, *The Diminishing Importance of Middle East Oil* (New York, NY: Holmes and Meier, 1982).
14. Penrose, op. cit.
15. P. Stevens, Jo*int Ventures in Middle East Oil, 1957–1975* (London: Graham and Trotman, 1976).
16. P. R. Odell, *Oil and World Power*, 1st to 8th Edition (London: Penguin Books, 1971 to 1986).
17. Cordesman, op. cit.
18. D. Hirst, *Oil and Public Opinion in the Middle East* (London: Faber and Faber, 1966).
19. R.F. Mahmoud and J.N. Beck, "Why the Middle East Fields may Produce Oil Forever," *Offshore*, April 1995, 56–62.

20. International Energy Agency, *World Energy Outlook to 2030* (Paris: OECD, 2002). See also IEA's *Energy to 2050: Scenarios for a Sustainable Future* (Paris: OECD/IEA, 2003) and IEA's *World Energy Investment Outlook* (Paris: OECD/IEA, 2003).

21. Based on P. R. Odell, *Why Carbon Fuels will Dominate the 21st Century's Global Energy Economy* (Brentwood: Multi-Science Publishing, 2004).

22. V. Smil, *Energy at the Crossroads, Global Perspectives and Uncertainties* (Cambridge, MA: MIT Press, 2003).

Bibliography

Adelman, M A. *The Economics of Petroleum Supply: Papers by M.A. Adelman 1962–1993* (Cambridge, MA: MIT Press, 1993).

Adelman, M. A. *The Genie Out of the Bottle: World Oil since 1970* (Cambridge, MA: MIT Press, 1995).

Adelman, M. A. *The World Petroleum Market* (Baltimore, MA: Johns Hopkins University Press, 1972).

Adelman, M.A. and M.C. Lynch. *Natural Gas Supply to 2100* (Hoersholm, Denmark: IGU, 2003).

Alexander, Lewis. "Overview: Inflation and Monetary Policy." *Global Economic Outlook and Strategy* (2006).

Alhajji, A.F. "The Impact of Dualism on the GCC Private Sector?" Oil, *Gas and Energy Law Intelligence* (OGEL) vol. 3, no. 2 (June 2005).

Alhajji, A.F. "What Have we Learnt from the Experience of Low Oil Prices? *OPEC Review* (September 2001).

AME Info. "Mid-East IPO obsession raises over USD 6 billion in 2005." Report on the first Middle East IPO Summit, Dubai in AME Info, February 25, 2006 (http://www.ameinfo.com/78222.html).

Anderson, D. *Energy and Economic Prosperity* (New York, NY: U.N. Development Program, 2001).

Arad, Ruth W., et al., *Sharing Global Resources* (New York, NY: Council on Foreign Relations, 1979)

Barkindo, Mohammed. "Energy Supply and Demand Security," EUROPIA Conference, London, February 15–16, 2006 (http://www.opec.org/opecna/Speeches/2006/Europia.htm).

Bernanke, B.S., M. Gertler and M. Watson. "Systematic Monetary Policy and the Effects of Oil Price Shocks." *Brookings Papers on Economic Activity*, 1997.

Bignell, Ray. "Global Refining Developments and Investment Trends" *Oxford Energy Forum*, Issue 44 (February 2004).

Blanch, Francisco G. "The Sustainability of Higher Oil Prices, The Revenge of the Old Economy" (Goldman Sachs, February 2005).

BP Statistical Review of World Energy (London: BP, June, 2005).

BP Statistical Review of the World Oil Industry (London: BP, 1955 to1980)

BP Statistical Review of World Energy (London: BP, various years)

BP/Amoco Statistical Review of World Energy (London: BP/Amoco, 2000).

Brossard, Emma B. *Petroleum: Politics and Power* (Tulsa, OK: Pennwell Books, 1983).

Brynen, Rex. "Economic Crisis and Post-Rentier Democratization in the Arab World: The Case of Jordan." *Canadian Journal of Political Science* vol. 25, no. 1 (1992).

China National Bureau of Statistics (CNBS). *Statistical Yearbook of China* (Beijing: CNBS, 2005).

Cordesman, Anthony H. *Energy Developments in the Middle East* (Westport, CT: Praeger Publishers, 2004).

Cordesman, Anthony H. "Energy Developments in the Middle East." *The Journal of Energy Literature* vol. 11, no. 1 (June 2005).

Cremer, Jacques and Djavad Salehi-Isfahani. *Models of the Oil Market* (Amsterdam: Harwood Academic Publishers, 1991).

Daley, George, James M. Griffin and Henry B. Steele. "Recent Oil Price Escalations: Implications for OPEC Stability," in James M. Griffin and David J. Teece (eds), *OPEC Behavior and World Oil Prices* (London: George Allen & Unwin, 1982).

Doran, Charles F. "Economics and Security in the Gulf," in David E. Long and Christian Koch (eds) *Gulf Security in the Twenty-First Century* (Abu Dhabi: The Emirates Center for Strategic Studies and Research, 1997).

Ehrhardt, Franz. "Refining and Price: Core Issues, Challenges and Opportunities." *Oxford Energy Forum*, no. 62 (August 2005).

Energy Information Administration (EIA). *Short-Term Energy Outlook* (Washington, DC: USDOE, October 2005).

Energy Information Administration. "The Impact of Environmental Compliance Costs on US Refining Profitability, 1995–2001" (Washington, DC: EIA, May 2003).

Energy Information Administration. *Annual Energy Outlook 2005 with Projections to 2025*, February 2005, Washington, DC (http://www.eia.doe.gov/oiaf/aeo/).

Energy Information Administration. *International Energy Outlook 2005*, July 2005, Washington, DC (http://www.eia.doe.gov/oiaf/ieo/index.html/)

Energy Information Administration. *Petroleum 1996: Issues and Trends* (Washington, DC: EIA, 1996).

Energy Information Administration. *The Global Liquefied Natural Gas Market: Status and Outlook* (Washington, DC: EIA, December 2003).

Energy Intelligence Research. *The International Crude Oil Market Handbook, 2006* (Energy Intelligence, Fifth Edition, December 2005).

ENI World Oil and Gas Review 2006 (Rome: ENI, 2006).

European Commission. *Towards a Hydrogen Based Economy* (Brussels: European Commission, 2003).

FACTS Inc. *Asia-Pacific Databook 1: Supply, Demand & Prices* (Honolulu, HI: FACTS, Fall 2005).

FACTS Inc. *Asia-Pacific Databook 2: Refinery Configuration & Construction* (Honolulu, HI: FACTS, Fall 2005).

FACTS Inc. *Asia-Pacific Databook 3: Oil Product Balances* (Honolulu, HI: FACTS, Fall 2005).

FACTS Inc. *Middle East Petroleum Databook* (Honolulu, HI: FACTS, Fall 2005).

Fasano, Ugo and Qing Wang. "Fiscal Expenditure Policy and Non-Oil Economic Growth: Evidence from GCC Countries." *IMF Working Paper* 01/195 (Washington, DC: International Monetary Fund, 2001).

Fasano, Ugo and Qing Wang. "Testing the Relationship between Government Spending and Revenue: Evidence from GCC Countries." *IMF Working Paper* 02/201 (Washington, DC: International Monetary Fund, 2002).

Fasano, Ugo and Zubair Iqbal. "GCC Countries: From Oil Dependence to Diversification" (Washington, DC: International Monetary Fund, 2003).

Fasano, Ugo et al. "Monetary Union among Member Countries of the Gulf Cooperation Council," *IMF Occasional Paper* no. 221 (Washington, DC: International Monetary Fund, 2003).

Fasano, Ugo. "Review of the Experience with Oil Stabilization and Savings Funds in Selected Countries." *IMF Working Paper* 00/112 (Washington, DC: International Monetary Fund, 2000).

Fawzi, Aloulou. "Gas to Liquids: A New Frontier for Natural Gas." *International Energy Outlook 2005* (Washington, DC: EIA, July 2005) (http:// www.eia.doe.gov/oiaf/ieo/index.html/).

Fawzi, Aloulou. Personal interview with H.E. Abdullah bin Hamad Al Attiyah, Second Deputy Premier and Minister of Energy and Industry of the State of Qatar, December 2003, Washington, DC.

Fawzi, Aloulou. Personal interview with Dr. Ibrahim B. Ibrahim, Chairman of Marketing and Vice Chairman of the Board of Qatar RasGas Company, December 18, 2003, Washington, DC.

Fawzi, Aloulou. Personal interview with Idriss Aljazairi, Algerian Ambassador to the United States, December 24, 2003, Washington, DC.

Fawzi, Aloulou. Personal interview with James T. Jensen, President of Jensen Associates Inc., January 7, 2004, Weston, Massachusetts (http://JAI-Energy.com).

Fawzi, Aloulou. Personal interview with Khaldoon Al Mubarak, Executive Vice President-Corporate of Dolphin Energy Limited at the USDOE's LNG Ministerial Summit, December 18, 2003, Washington, DC.

Francisco Parra, *Oil Politics: A Modern History of Petroleum* (New York, NY: IB Tauris, 2004).

Fuller, Graham E. and Ian O. Lesser. "Persian Gulf Myths." *Foreign Affairs*, vol. 76, no. 3 (May/June 1997).

Gately, Dermot. "OPEC's Incentives for Faster Output Growth." *Energy Journal*, vol. 25, no. 2 (September 2004).

Gause, Gregory. *Oil Monarchies: Domestic and Security Challenges in the Arab Gulf* (New York, NY: Council on Foreign Relations, 1994).

Gelb, A.H. and Associates. *Windfall Gains: Blessing or Curse?* (New York, NY: Oxford University Press, 1988).

Girgis, M., F. Hadad-Zervose and A. Coulibaly. "A Strategy for Sustainable Employment for GCC Nationals," *Working Paper*, World Bank (2003).

Goldstein, Andrea, et al. *The Rise of China and India: What's in it for Africa* (Paris: OECD/ Development Centre Studies, 2006).

Goodwin, Kevin. "The State of the Refining Industry" *Oxford Energy Forum*, Issue 44 (February 2004).

Gordon, Richard et al. *World Petroleum Markets: A Framework for Reliable Projections* (Washington, DC: World Bank Technical Paper no. 92, Industry and Energy Series, 1988).

Granger, C.W.J. “Some Recent Developments in a Concept of Causality.” *Journal of Econometrics* vol. 39, 199–211 (1988).

Griffin, James M. and Henry B. Steele. *Energy Economics and Policy*, second edition (Orlando, FL: Academic Press, 1986).

Gulf Research Center. *Gulf Yearbook, 2005–2006* (Dubai: GRC, 2006).

Hall, Marshall. “Refining Margins and Investment.” *Oxford Energy Forum*, Issue 62 (August 2005).

Hamilton, James. “Oil and the Macroeconomy since World War II.” *Journal of Political Economy*, vol. 91, no. 2 (1983).

Harik, I. and D. J. Sullivan. *Privatization and Liberalization in the Middle East* (Bloomington IN: Indiana University Press, 1992).

Helm, Dieter, John Kay and David Thompson. “Energy Policy and the Role of the State in the Market for Energy.” *Fiscal Studies* vol. 9, no. 1 (February 1988).

Hirst, D. *Oil and Public Opinion in the Middle East* (London: Faber and Faber, 1966).

Hooker, Mark A. “Are Oil Shocks Inflationary? Asymmetric and Nonlinear Specifications versus Changes in Regime.” *Finance and Economics Discussion Series* 1999–65 (Washington, DC: Board of Governors of the Federal Reserve System, 1999).

Horsnell Paul and Robert Mabro. *Oil Markets and Prices: The Brent Market and the Formation of World Oil Prices* (Oxford: Oxford University Press, 1997)

Ineson, Robert. “The Incoming Tide: LNG Surges into North America.” Decision Brief (Cambridge, MA: CERA, January, 2004).

Institut Français du Pétrole. *Oil and Gas Exploration and Production, Reserves, Costs, Contracts* (Paris: Centre for Economics and Management, IFP-School, 2004).

Institute for International Finance (IIF). Working Group on Governance and Transparency (www.iif.com).

International Energy Agency. *Energy to 2050: Scenarios for a Sustainable Future* (Paris: OECD/IEA, 2003).

International Energy Agency. *Oil Market Report* (Paris: IEA/OECD, October, 2005).

International Energy Agency. *World Energy Investment Outlook 2003* (Paris: IEA/OECD, 2003).

International Energy Agency. *World Energy Outlook 2004* (Paris: OECD/IEA, 2004).

International Energy Agency. *World Energy Outlook 2005* (Paris: OECD/IEA, 2005).

International Energy Agency. *World Energy Outlook to 2030* (Paris: OECD, 2002).

International Monetary Fund (IMF). *Oil Market Developments and Issues* (Washington, DC: IMF, 2005)

International Monetary Fund. *World Economic Outlook, Globalization and External Imbalances* (Washington, DC: IMF, April 2005).

Issawi, C. and M. Yeganeh. *The Economics of Middle Eastern Oil* (New York, NY: Praeger Publishers, 1962).

Jaffe Amy M. and Ronald Soligo. "A Note on Saudi Price Discrimination." *Energy Journal*, vol. 21, no. 1 (January 2000).

Jaffe, Amy M. and Edward L. Morse. "OPEC in Confrontation with Globalization," in Jan H. Kalicki and David L. Goldwyn (eds) *Energy and Security* (Baltimore, MD: The Johns Hopkins University Press, 2005).

Johansen, S. "Statistical Methods of Econometrics." *Journal of Economic Dynamics and Control* 12 (1988).

Kahler, Miles. "Economic Security in an Era of Globalization: Definition and Provision." *The Pacific Review* vol. 17, no. 4 (2004).

Kanovsky, E. *The Diminishing Importance of Middle East Oil* (New York, NY: Holmes and Meier, 1982).

Karawan, Ibrahim. "The Erosion of Consensus: Perceptions of GCC States of a Changing Region," in Lawrence Potter and Gary Sick (eds) *Security in the Persian Gulf: Origins, Obstacles, and the Search for Consensus* (New York, NY: Palgrave, 2002).

Krasner, Steven D. "A Statist Interpretation of American Oil Policy and the Middle East." *Political Science Quarterly*, vol. 94, no.1 (Spring 1979).

Larson Alan P. and David M. Marchick, *Foreign Investment and National Security: Getting the Balance Right* (New York, NY: Council on Foreign Relations, 2006).

Lenczowski, G. *Oil and State in the Middle East* (Ithaca, NY: Cornell University Press, 1960).

Longrigg, S. H. *Oil in the Middle East*, 3rd Edition (Oxford: Oxford University Press, 1968).

Lopez-Claros, Augusto and Klaus Schwab (eds) *The Arab Competitiveness Report* (Switzerland: Palgrave Macmillan, 2005)

Lubell, H. *Middle East Oil Crises and W. Europe's Energy Supplies* (New York, NY: Elsevier Press, 1971).

Mahmoud R.F. and J.N. Beck: "Why the Middle East Fields may Produce Oil Forever." *Offshore*, April 1995.

Mandelbaum, Michael. *The Case for Goliath: How America Acts as the World's Government in the Twenty-first Century* (New York, NY: Public Affairs Books, 2006).

Moran, Theodore. "Modelling OPEC Behavior: Economic and Politic Alternatives," in James M. Griffin and David J. Teece (eds), *OPEC Behavior and World Oil Prices* (London: George Allen & Unwin, 1982).

Morse, Edward L. "Energy Breaks the Economic Rules." *SAISPHERE* 2005 (January 2006).

Morse, Edward L. and Stenvoll Thomas. "The Refinery Capacity Treadmill." Hess Energy Trading Company, LLC, *Weekly Market View*, May 3, 2005.

Nesadurai, Helen E.S. "Introduction: Economic Security, Globalization and Governance." *The Pacific Review* vol. 17, no. 4 (2004).

New York Mercantile Exchange. *Crack Spread Handbook* (New York, NY: NYMEX, 2000).

Noreng, Øystein. *Oil and Islam* (Chichester: Wiley and Sons, 1998).

O'Brien John B. and Jensen Scott. "New Formula Yields Coking Refinery Margins more Reliably." *Oil & Gas Journal*, May 16, 2005.

Odell, P. R. "The Global Oil Industry—The Location of Production." *Regional Studies*, vol. 31, no. 3 (1997).

Odell, P. R. *Why Carbon Fuels will Dominate the 21st Century's Global Energy Economy* (Brentwood: Multi-Science Publishing, 2004).

Odell, P. R. *Oil and World Power*, 1st to 8th Edition (London: Penguin Books, 1971 to 1986).

Odell, P. R. "The Significance of Oil." *Journal of Contemporary History*, vol.3, no.3 (1968).

Odell, P. R. "World Resources, Reserves and Production." *Energy Journal*, Special Issue (1994).

Odell, P. R. and K.E. Rosing, *The Future of Oil: World Oil Resources and Use* (London: Kogan Page, 1983).

Odell, P. R. *Oil and Gas: Crises and Controversies, 1961–2000* (vol. 1), *Global Issues* (Brentwood: Multi-Science Publishing, 2001.

Oil and Gas Journal, vol. 102, no. 47 (December 20, 2004).

OPEC Secretariat. *Oil Outlook to 2025* (Vienna: OPEC Secretariat paper, 10th International Energy Forum, Doha, April 22–24, 2006).

Oppenheim, V.H. "Why Oil Prices Go Up: The Past: We Pushed Them. *Foreign Policy*, no. 25 (Winter 1976–77).

Organization of Petroleum Exporting Countries (OPEC). *Monthly Oil Market Report* (Vienna: October, 2005).

Penrose, E.T., *The Large International Firm in Developing Countries* (London: Allen and Unwin, 1968).

Peter Schweizer, *Victory: The Reagan Administration's Secret Strategy that Hastened the Collapse of the Soviet Union* (New York, NY: Atlantic Monthly Press, 1994).

Petroleum Industry Research Associates. *Scenario Planning Service: Annual Guidebook 2005* (New York, NY: PIRA, 2005).

Petroleum Industry Research Associates. *Scenario Planning Service: Annual Guidebook 2006* (New York, NY: PIRA, 2006).

Rodrik, Dani. *Has Globalization Gone Too Far?* (Washington: Institute for International Economics, 1997).

Roger Hansen (ed). *Rich and Poor Nations in the World Economy* (New York, NY: Council on Foreign Relations, 1978).

Rogers, David. "US Annual War Spending Grows." *Wall Street Journal* (March 8, 2006).

Sachs, Jeffrey and Andrew Warner. "Natural Resource Abundance and Economic Growth." Harvard Institute for International Development Discussion Paper (1995).

Seymour, Ian. *OPEC: Instrument of Change* (New York, NY: St. Martin's Press, 1981)

Shell International Ltd. *Energy Needs, Choices and Possibilities: Scenarios to 2050* (London: Global Business Environment, 2001).

Shwadran, B. *The Middle East, Oil and the Great Powers*, 3rd Edition (New York, NY: Wiley, 1979).

Skeet, Ian. *OPEC, Twenty-Five Years of Prices and Politics* (Cambridge: Cambridge Energy Studies, 1988).

Smil, V. *Energy at the Crossroads, Global Perspectives and Uncertainties* (Cambridge, MA: MIT Press, 2003).

Stevens, P. *Joint Ventures in Middle East Oil, 1957–1975* (London: Graham and Trotman, 1976).

Stevens, Paul. *Resource Impact – Curse or Blessing? A Literature Survey* London, International Petroleum Industry Environmental Conservation Association, 2003 (www.ipieca.org/downloads/social/PStevens_resourceimpact_final.doc).

Stevens, Paul. *The Economics of Energy* vol. I and II (Cheltenham, UK: The International Library of Critical Writings in Economics, 2000).

Stiglitz, Joseph. *Globalization and its Discontents* (New York, NY: W.W. Norton & Company, 2002).

Stocking, G. W. *Middle East Oil: A Study in Political and Economic Controversy* (London: Allen Lane, 1971).

Telhami, Shibley. "The Persian Gulf: Understanding the American Oil Strategy." *The Brookings Review*, vol. 20, no.2 (Spring 2002).

Terreson, Douglas. "The Golden Age of Refining." *Oxford Energy Forum*, Issue 62 (August 2005).

Qatar: The Modern State (Washington, DC: Qatar embassy, November 2004).

United States Geological Survey. *World Petroleum Assessment* (Reston, VA: USGS, 2000).

Vandewalle, Dirk. "Social Contracts, Institutional Development and Economic Growth and Reform in Middle East Oil Exporters." Unpublished Paper, Dartmouth College (2003).

Verleger, Philip K. "The Evolution of Oil as a Commodity," in Richard L. Gordon, et al., (eds), *Energy: Markets and Regulation* (Cambridge, MA: MIT Press, 1987).

Wauquier, J.-P. *El Refino del Petróleo* (España: Fundación Repsol YPF, Instituto Superior de la Energía, 2004).

World Bank. *Better Governance for Development in the Middle East and North Africa: Enhancing Inclusiveness and Accountability* (Washington, DC: World Bank, 2003).

World Bank. *Claiming the Future: Choosing Prosperity in the Middle East and North Africa* (Washington, DC: World Bank, 1995).

World Bank. *Economic Developments and Prospects: Oil Booms and Revenues Management* (Washington, DC: World Bank, 2005).

World Bank. *Gender and Development in the Middle East and North Africa: Women in the Public Sphere* (Washington, DC: World Bank, 2004).

World Bank. *Jobs, Growth and Governance in the Middle East and North Africa—Unlocking the Potential for Prosperity* (Washington, DC: World Bank, 2003).

World Bank. *Trade, Investment and Development in the Middle East and North Africa—Engaging with the World* (Washington, DC: World Bank, 2003).

World Bank. *Unlocking the Employment Potential in the Middle East and North Africa—Toward a New Social Contract* (Washington, DC: World Bank, 2004).

World Bank. *World Development Report* (Washington, DC: World Bank, 2005).

World Market Research Center. Country Reports. London, September 2005 (www.wmrc.com).

World Petroleum Congress. *New Hydrocarbon Provinces of the 21st Century*, vol. 2 (London: WPC, 2002).

Wu, Kang and Fereidun Fesharaki. "As Oil Demand Surges, China Adds and Expands Refineries." *Oil and Gas Journal* vol. 103, no. 28 (July 25, 2005).

Yergin, Daniel and Michael Stoppard. "The Next Prize." *Foreign Affairs*, November/December 2003.

Yousef, Tarik. "Development, Growth and Policy Reform in the Middle East and North Africa since 1950." *Journal of Economic Perspectives* vol. 18, no. 3 (2004).

Yousef, Tarik. "Structural Reforms, the Investment Climate and Private Sector Development in the Arab World," in Augusto Lopez-Claros and Klaus Schwab (eds) *The Arab Competitiveness Report* (Switzerland: Palgrave Macmillan, 2005).

Zanoyan, Vahan. "After the Oil Boom: The Holiday Ends in the Gulf." *Foreign Affairs* vol. 74, no. 6 (1997).

INDEX

P